KW-479-847

# Women and Farming

## Property and Power

Sally Shortall
*Reader*
*Department of Sociology and Social Policy*
*Queen's University of Belfast*

Consultant Editor: Jo Campling

UNIVERSITY OF PLYMOUTH
SEALE HAYNE
LIBRARY

First published in Great Britain 1999 by
**MACMILLAN PRESS LTD**
Houndmills, Basingstoke, Hampshire RG21 6XS and London
Companies and representatives throughout the world

A catalogue record for this book is available from the British Library.

ISBN 0–333–66465–5 hardcover
ISBN 0–333–66466–3 paperback

First published in the United States of America 1999 by
**ST. MARTIN'S PRESS, INC.,**
Scholarly and Reference Division,
175 Fifth Avenue, New York, N.Y. 10010

ISBN 0–312–21990–3

Library of Congress Cataloging-in-Publication Data
Shortall, S.
Women and farming : property and power / Sally Shortall ;
consultant editor Jo Campling.
p.   cm.
Includes bibliographical references and index.
ISBN 0–312–21990–3 (cloth)
1. Women farmers.   2. Women in agriculture.   3. Family farms.
4. Farm ownership.   5. Property.   6. Dairy farmers—History—19th
century.   7. Sex discrimination against women.   I. Campling, Jo.
II. Title.
HD6073.F3S53   1999
332'.042—dc21                                                          98–44285
                                                                              CIP

© Sally Shortall 1999

All rights reserved. No reproduction, copy or transmission of this publication may be made
without written permission.

No paragraph of this publication may be reproduced, copied or transmitted save with
written permission or in accordance with the provisions of the Copyright, Designs and
Patents Act 1988, or under the terms of any licence permitting limited copying issued by
the Copyright Licensing Agency, 90 Tottenham Court Road, London W1P 9HE.

Any person who does any unauthorised act in relation to this publication may be liable to
criminal prosecution and civil claims for damages.

The author has asserted her right to be identified as the author of this work in accordance
with the Copyright, Designs and Patents Act 1988.

This book is printed on paper suitable for recycling and made from fully managed and
sustained forest sources.

10   9   8   7   6   5   4   3   2   1
08   07   06   05   04   03   02   01   00   99

Printed and bound in Great Britain by
Antony Rowe Ltd, Chippenham, Wiltshire

For Manos

# Contents

# Acknowledgements

I am grateful to many colleagues and friends who have commented on drafts of this book. Calla Rowan has always shown an interest in my work, and her assistance in keeping me updated on the Canadian Farm Women's Network was invaluable. Within my department, Kate Mulholland and Sam Porter generously gave of their time to read and comment on earlier drafts of the book.

I particularly want to thank Chris Curtin who, on a very tight deadline gave me very useful comments. Recognising my fear that my babies would be born before the book was published, he sent thoughts and references in emails titled 'books and babies'. Mark Shucksmith also made many useful comments.

The Nuffield Foundation funded my research in the North of Ireland, and the Canadian Rural Restructuring Foundation (at that time, the Agricultural and Rural Restructuring Group) funded my research in Newfoundland. I am grateful to both organisations.

Jo Campling provided invaluable advice and guidance throughout the project. In the early stages in particular, her good humour and advice were very important.

Finally, I would like to thank Manos Marangudakis. As always he was my most ardent critic, and my greatest source of support. Without his help and encouragement, I might never have finished the book. It is dedicated to him.

# 1   Introduction

Family farming is the prevalent system of farming in western Europe and North America. There is a two-way relationship between the farm family and the farm business. The family is affected by virtue of living on and operating a business, and the business is affected by the fact that it is operated by a family. The classic definition of the family farm is that it is a unit that combines the four factors of production: land, capital, management and labour. Typically, ownership is combined with management control; the family lives on the farm, and ownership and management control are transferred between the generations with the passage of time (Gasson and Errington, 1993).

Various aspects of the relationship between the farm family and the farm business have been the subject of research. Studies of the farm-family business have considered whether or not family ownership and control of the business enhances or diminishes business performance. The composition of the family and the future transmission of the farm are thought to affect risk-taking, investment and expansion of the farm. The combination of management and control of the farm business, and the intergenerational transfer of the farm business, are almost always organised along gender lines. Men own and control farms, and men transfer farms to their sons. The farm-family business is not only a relationship between the farm family and the farm business. It also embodies a whole set of relationships within the farm family. In many ways an understanding of the position of women in farming is often complicated by the discourse of the 'family farm'. It focuses attention on the family unit rather than the inter-personal, economic and social status positions of those within the family. It is women's entry to and position within the farm family that is the starting point for the discussion of power and property that is the subject of this book.

Early feminist research sought to make the role of women on farms 'visible'.[1] As the title of Sachs' (1983) well-known study suggested, women were, and many would argue still are, the 'invisible farmers'. The traditional definitions of farm work focus on the work of the owner and manager, and frequently the work of women on farms goes unnamed and unrecognised. Women rarely inherit land. Their typical entry to farming, and to the farm family, is through marriage. Women's whole relationship to farming is shaped by their route of entry and position within the farm family. It not only affects inter-personal relationships within the family, but also women's role in the public space of farming. Women are under-represented in farming organisations, in training programmes, and in the politics of farming.

My central argument in this book is that in order to understand the position of farm women, we must look at how power is organ-ised in farming. Farming culture affords men more power than women. Central to this is access to property. Land is transferred inter-generationally within families. It is a transaction mediated by cultural norms, and land is typically passed from father to son. Acquisition of land is based on sex, and we see the very foundation of different positions for men and women on farms being formed. Men constitute the constant family line through which land is passed, and women float in and out. Not having independent access to land shapes women's position within the family and within the public space of farming.

There are two main expressions of power that I will consider in this book. First, I will examine the power manifest in what Lukes (1974) called the 'taken for granted'. This refers to a body of customs, beliefs and social practices that are accepted without question. The solidity of these customary practices rests on their legitimacy, which places them beyond question. The most obvious example of this form of power that I will discuss in the book is the inter-generational transfer of land from father to son. De Haan (1994) wonders why the acquisi-tion of wealth and status by pure accident of birth has not been of greater sociological interest. Pure accident of sex is equally part of the inheritance of farms, and it is the legitimacy of the social practices regulating land transfer that place it beyond question.

The second expression of power I will examine, and it is related to the first, is control of resources. Property is the central economic resource in farming, and its acquisition is regulated by the customs and traditions described above. In other words, women have limited access to land, and thus to the central economic resource in farming. Property is not only an economic resource, and I will discuss the status and prestige attached to land ownership. Nor is property the only resource in farming, although I will argue that it provides easy access to the other central resources. Access to agricultural know-ledge, organisational resources, ideological and political power are all connected to land ownership.

To talk of power is to imply the possibility for change. The validity claims on which an expression of power rests are always open to question. Many examples of change will arise throughout the book. Change, as we will see, can occur from a number of angles. Change in the situation of farm women can occur because of more general changes in gender ideology, or in the values and ethos of society. Equally, organised protest can lead to change.

Chapters 2 and 3 consider power and property respectively. Chapter 2 outlines the concept of power central to the book. I outline what is the source of power in farming, and how it is identified. Lukes' theory of power is central to this chapter as it allows us consider power in situations where there appears to be consent, and where a given situation is taken for granted. Property is considered separately in Chapter 3. Property and power are intrinsically connected in farming. Property is an economic resource, and access to it is regulated by customs, beliefs and social practices that are accepted as legitimate. The blatant gender discrimination they embody goes relatively unquestioned for this reason. Revisiting the work of John Locke allows us to consider the source of the legitimacy of individual private property ownership, of transfer rights and inheritance and the protection of these processes by the state. All of these are central to women's position in farming: land ownership, inheritance, and the tacit protection of transfer practices by the state. The chapters on power and property provide the foundation for the rest of the book.

The following chapters consider a number of aspects of farming culture and farming practice and the role of women. How power and property shape the situation of women is central to this analysis. Chapters 4 and 5 are both historical chapters. Chapter 4 looks at women's relationship to property in Ireland throughout the twentieth century. Taking an historical perspective illustrates the tenacity of patrilineal land transfer, as well as the traditions and social customs that regulate the process. This chapter illustrates the considerable status and prestige, as well as economic power, that accompanies land ownership. Chapter 5 considers the commercialisation of the dairy industry that occurred in western Europe and North America in the late nineteenth and early twentieth century. I argue that women's disengagement from the dairy industry during this period was primarily the result of their relationship to property, and their lack of resources for bringing their grievances to public attention. Dairy women at this time had limited organisational resources, and limited political power. I document resistance by disaffected women, but argue that their limited access to public space, their primary identity as members of farm families, and most importantly, their lack of independent resources all militated against successful organisation.

The remaining chapters are contemporary. In Chapter 6 I turn to women's role in farming organisations. I look at three instances of women in farming organisations; firstly those few women in farming organisations which tend to be entirely male, secondly women on auxiliary committees within farming organisations, and thirdly, women's farming organisations. While women's farming organisations sound gendered in a way that farming organisations do not, this is only because the title of the latter belies their gendered composition. This chapter illustrates how organisational resources and power are tied to property ownership. By organising, women have increased their collective power. The Canadian Farm Women's Network is the clearest example of this enhanced power position. None the less, women are organising in a context that does not afford them sufficient resources to organise to maximum effect. They remain outside the political apparatus. The perception of farming as an individual male occupation persists in farming organisations.

Chapter 7 looks at agricultural education and training. Again we see the construction of agricultural education and knowledge that assumes an individual male farming operator. Training programmes are geared towards men, and relate to the work they do on farms. Women's entry to farming through marriage means that they rarely have any training prior to becoming part of a farm family. The social construction of agricultural knowledge, such that some areas are more valued than others, represents an exercise of power. This varied valuation of knowledge reflects and reinforces the stratified valuation of men and women's farm work. The resources and institutional apparatus available to impart productive agricultural knowledge are also an indication of power. In Chapter 7 we will see that the sources of funding and institutional apparatus to deliver training to women in the cases examined are tenuous.

Chapter 8 argues that the state is a powerful actor shaping the position of women in farming. Whether organised protest by farm women's groups remains outside the political apparatus or is incorporated into institutional channels is largely dependent on how the state is organised, and the wider political context in which women's groups are placed. Three examples are examined: Norway as an example of a social-democratic state, the North of Ireland as a male-breadwinner state, and Canada as a liberal-pluralist state. In each case the ethos and values of the state affects the lives of women on farms. Norway has directly legislated on land transfer, as the gendered access to farming was contrary to its deep commitment to gender equality. In the North of Ireland we see women forming auxiliary farm family committees within farming organisations, and specific provisions for women being organised for them as 'wives'. The idea of a farming male bread-winner underpins all of these developments. In Canada, a liberal-pluralist state which funds interest groups, farm women have formed something of a social movement, and lobby the state for change. The chapter on the state, in particular, illustrates how power is contested, and how change, in different forms, occurs in different social circumstances.

Most of the empirical examples used throughout the book refer to Ireland, North and South. However, I believe the concepts and issues raised are common to most women on farms in the western world. My disproportionate use of Ireland as an example stems from my familiarity with the Irish situation, and the fact that it is where I have conducted most of my research. Ireland is typical of family farming, displaying the most common characteristics. Comparisons are made with other countries, mostly Canada, where I have also carried out research, and Norway, which is such a distinct case that it makes a very useful point of comparison. My own research in the South of Ireland was conducted over a five-month period in 1987 and 1989.[2] The main focus of the study was to identify the various factors which render women's role in farming invisible, and why women continue to subscribe to such a situation. The main argument advanced was that an analysis of power is key to understanding the lack of a greater expression of grievance by women about a context that so clearly disadvantages them. It was a qualitative study, involving participant observation, several in-depth interviews with twenty women, married or widowed (two cases) and living on farms, as well as interviews with members of farm organisations, farm advisory services and local organisations. In the North of Ireland, twenty-seven women involved in farm women's training groups were interviewed in 1995. Ten other interviews were carried out. Women agricultural advisers instrumental in setting up the groups were interviewed, along with people in the agricultural college who had responsibility for the training groups.[3] Work with the Canadian Farm Women's Network was conducted between 1990 and 1992. This involved attending provincial meetings of farm women's groups and federal meetings of the farm women's network. Interviews were conducted with leaders of provincial networks and with farm women involved as members, and during

February 1990, a week was spent living and working with members of farm women's groups in Newfoundland. Part of this week was spent living with, working with and interviewing the president of the Canadian Farm Women's Network, who at the time was from Newfoundland.[4] While the Norwegian case is built from secondary analysis, this is justifiable on a number of grounds: firstly, the purpose of the Norwegian example is not to establish myself as an expert on Norwegian agriculture and the lives of Norwegian farm women, but rather to pen a broad illustrative case of a different political context to that presented by the other case studies. Norway provides an ideal example. Secondly, Norway has a very rich literature available in English on farm women and change, and on gender equality, social democracy and the state, of which I have availed myself.[5] Ireland North and South, Canada and Norway are the examples most widely used in the following chapters. Other examples are sometimes used, particularly in Chapter 5, when I look at changes in the dairy industry. The common themes of power and property underpin all the issues discussed and illustrated by the various case studies. I believe they have wider applicability than the cases examined in this book. Future research may show this to be the case.

A final point, but worth clarifying: what do I mean by farm women, or women on farms? Sometimes I have been asked if I am talking about women farming independently in their own right. I am not. There are very few such women. In general, I am referring to women who have entered farming following their marriage to men who owned land and were already farming. To refer to these women as farm wives, or farmers' wives, is dated and inappropriate for obvious reasons. Some colleagues and I spent an amusing hour one morning at the European Congress for Rural Sociology in 1997 trying to come up with gender-neutral alternatives to 'farmer' and 'farm woman'. It was fun, but we had little success.

# 2 Power

## INTRODUCTION

All of the following chapters are imbued with an analysis of power. As I have said, my argument is that the structure of farming culture affords men more power than women. Firstly, property ownership is a source of power. Owning land provides economic power, as well as varying degrees of social, cultural and political power. Secondly, there is power associated with customs and practices that hold traditional patterns of land transfer in place. It is the legitimacy of these customs that allows their persistence. Here lies the possibility for change. If the legitimacy of customs and practices is questioned, it is possible for change to occur. To speak of power implies the possibility for change.

Women in farming fare badly in both the aspects of power I have mentioned. Firstly, because they rarely own land, they have limited independent access to land as an economic resource, and the consequent status, prestige and political power land-ownership brings. Secondly, the legitimacy of traditional patterns of land transfer means that women's disinherited position is relatively unquestioned, although an example to the contrary will be discussed in the following chapters.

The rest of the book considers how power is organised in farming culture, and its different implications for men and women. Our accepted understanding of private property, discussed in the next chapter, bestows economic and social power and is upheld by the state. An examination of customary practices governing the transfer of land in Ireland demonstrates the power of tradition, and the power associated with property ownership. Owning land provides additional assets, including political power and organisational resources. These are considered in the chapter on dairying in the late nineteenth century, where women's lack of resources prevented a more successful organisation of protest at the changes that occurred in the industry. They are also considered in the chapter on organisations, where I will examine contemporary forms of women's organisations. The discussion of power that runs throughout the book includes an examination of change in the structure of power in farming. The last chapter most overtly considers change, and the various forms it can take. However, the question of change also runs throughout the other chapters. When I look at women and property in twentieth-century Ireland, it is clear that zero-sum expressions of power associated with land ownership have changed.

The concept of power is central to this book. It is imperative then to spell out what I mean by power, and how it will be used. As we all know, power is a messy and contentious concept. In this chapter, I will outline the understanding of power that runs through and connects the following chapters, and I will also outline what I consider to be the source of power in farming.

In order to disentangle the debate about power, it is useful to think of it as conducted on two levels. Firstly, there is a discussion about the sources of social power. This question has concerned theorists such as Marx (1971), Weber (1978), Parsons (1960), and more recently, Mann (1986; 1993). The second level on which the debate is conducted relates to a taxonomy of power. In other words, the debate centres around how to identify power and how it is exercised, as opposed to considering the sources of power. While some of the theorists who consider the sources of power also consider this aspect, others such as C. Wright Mills (1956), Bachrach and Baratz (1962), Lukes (1974) and Foucault (1980), have focused more exclusively on how an exercise of power can be identified, and what are the most far reaching and effective types of power. Interestingly, most of the debate on power has focused on this aspect, and consequently, definitions of what power actually is are hard to find. Any study of power needs to consider both these elements; firstly the source, and secondly how power is exercised and identified.

I will begin this chapter by looking at some of the debate on the meaning of power, and its source. My central argument throughout the book is that the primary source of power in farming is access to property. Property ownership in turn provides status and prestige, itself a source of power, and it also provides access to resources and the means to organise in order to obtain political power. Power in farming is visible in the organisational resources of farmers, their political influence, their ability to determine their resources, and their access to agricultural knowledge. As I will discuss in later chapters, all of these are connected to property ownership.

Access to property is socially constructed, and discriminates on the basis of sex. Men have greater access to property than women. Given that property and sex are central to this argument, I will discuss Marxism and feminism, since one would expect their theories on power to be particularly relevant. Both perspectives are taken up again in the next chapter, which deals more specifically with property. Alas, while both are useful, neither on their own take us far enough. The position of women within the farm family provides a peculiar nexus for theories of power. I will examine how functionalists, Marxists, feminists and Weberians can all offer explanations of the power arrangements within farming. It is the Weberian perspective, though, that I argue offers the most comprehensive analysis. Next I will consider the many 'faces' of power, or how it is exercised. Finally I will return more specifically to women in farming, and how these two levels of the power debate can be tied to an analysis of their lives.

## POWER AND ITS SOURCE

### Definitions

There is no standard, accepted definition of power. Weber has defined power as 'the chance of a man [*sic*] or a number of men to realise their own will in a social action even against the resistance of others who are participating in the action' (1978, p. 926). Power is an aspect of social relationships. An individual or group does not hold power in isolation, they hold it in relation to others. Weber's definition portrays power as a zero-sum game: in order for one person to obtain power, another loses some. It is also an individualistic perspective – power is possessed by individuals.

It was Parsons who developed the idea of collective power, whereby people through co-operation can enhance their joint power (Parsons, 1960). This idea of collective power is further developed by Mann (1986), who incorporates Weber and Parsons into one scheme. There are two dimensions to power: the Weberian (zero-sum) power and the Parsonian collective (created) power. According to Mann, each and every group of people is locked into both simultaneously. Collective power leads Mann to discuss organisational power, and 'organisational outflanking'. This results from the lack of collective organisation, which means that people are embedded within collective and distributive power organisations controlled by others, thus obviating their ability to question the exercise of power.

Mann, like Parsons, defines power as a 'generalised means' to attain whatever goals one wants to achieve. Power provides the means and ability to pursue and attain goals. Almost all theorists of power agree that while power may be maintained through coercion, the most stable form of power relationship is that where the arrangement is considered legitimate by all participating individuals (Giddens, 1971). According to Stinchcombe (1968), a power is reckoned legitimate whenever its holder can effectively call on other centres of power for support. It is thus the legitimacy of the claim in other centres of power which is crucial, and not its legitimacy among people who must take the consequences. This is one avenue of change in power relationships. As Habermas (1973) points out, legitimacy is a contestable validity claim. It is always open to question, and the changes we have witnessed throughout time are testimony to this fact. As we will see, so it is too for farm women. In Norway, the allodial law questions the legitimacy of a social custom that favours men as heirs. In the North of Ireland, *ad hoc* training provision for women represents an inchoate questioning of the legitimacy of a training service that only trains men.

In this book, I understand power as a generalised means, which can be used to maintain a situation, or to achieve change, to the selective advantage of a small or a broad social group. Power has individual and collective aspects, and one chapter in this book deals specifically with organisations. The stability of power relationships depends on their legitimacy, with social norms, customs and traditions, underpinned by legal support, reinforcing that legitimacy. I agree with Lukes (1974) that the very discussion of power relationships implies the possibility of change. The main types of change examined in this book are the result of the questioning of legitimacy claims of particular arrangements, and they take two overt forms. Firstly, there is change from 'above', that is, changes introduced by enlightened corporatist social-democratic states such as we see happening in Norway, where the transfer of land from father to son has been legally questioned. The second main type of change is from 'below'. Here we see *ad hoc* struggles to change the situation of farm women by individuals in organisations, as in the case in the North of Ireland, and other more organised forms of collective action where women come together to express grievances and organise alternative structures, as is the case in Canada, which again signals a questioning of the legitimacy of current arrangements. Let us turn now to consider the sources of social power.

## Marxists, Feminists, and Power

My reason for dealing with Marxism and feminism together in this sub-section is that, ostensibly, women on farms seem to offer an instance where the theories of both groups overlap. The different positions of men and women on farms is tied to ownership of private property. These different positions are socially constructed on the basis of sex. It would seem then that the two theories merge: the source of power is ownership of the means of production, and the owners are men.

*Marxists* stress relations to the means of production as the source of power. In other words, those who own the means of production have access to economic power. For Marxists there is a single source of power: economics. Political and ideological power emanate from economic power. The state operates to protect the rights of property owners, and their disproportionate share of the wealth produced by their employees. The Marxist theory of power is fundamentally one which focuses on inequality.

The same is true of *feminist* theories of power; they focus on inequality between the sexes. Marx, however, did not give a great deal of thought to inequalities between the sexes (Barrett, 1983). The source and solution of women's oppression fits neatly with the more general Marxist theory; Engels, who deals more comprehensively with the issue than Marx, identifies the source of women's oppression as the appropriation of private property by their husbands. The conflict between the sexes develops with the appearance of private ownership of wealth, and the reconciliation of the sexes is only possible when private property has been abolished. Thus, the fortunes of women and of oppressed classes are intimately connected; neither can be free until the basis of private property ownership has been abolished (Engels, 1973; Delmar, 1979).

While Marxist and feminist theories of power both focus on inequality, there has always been an uneasy relationship between the two perspectives. Marxist feminists have tried to marry the two positions, but it has not been an easy task. Feminism is based principally on a philosophy of egalitarianism. Underlying many feminist positions is a demand for equal rights for all individuals, and it is couched in terms of an appeal to moral rights and justice. Marxists on the other hand, claim to provide a scientific account of the exploitation we experience, with a view to overthrowing it (Barrett, 1983, although see Firestone, 1971). In other words, while both theories focus on inequalities of power, they identify different sources, and ultimately, different solutions. Feminists identify sex as the basis of power. Despite this powerful statement, feminist theories of power remain largely underdeveloped, and primarily descriptive (Barker and

Roberts, 1993; Pollert, 1996). Feminist theorists have, however, raised a number of difficult questions for Marxists. Feminists have argued that to analyse society exclusively in class terms ignores the distinctive social experiences of the sexes. Marxist demands, it is claimed, could be, and sometimes have, been satisfied without altering women's inequality to men. What does it mean for class analysis if it can be asserted that a social group is defined and exploited through means largely independent of the organisation of production? What if it could be demonstrated that capitalism would not be materially altered if the means of production were controlled by women (MacKinnon, 1982)?

One of the examples I will use in this book poses exactly this question for Marxists; how can they accommodate an instance where women establish equal rights to private property? In such a case, private property is maintained, it is simply the sex of the owners that has altered. The Allodial Law in Norway, introduced in 1974, made the eldest child the legal heir to the farm, regardless of gender. In other words, the introduction of this law furthers gender equality by providing women with increased opportunity to own private property. It does not question the capitalist arrangement, however, and in introducing this policy, the Norwegian state has not undermined rights to private property in any way. The difficulty for Marxist theorists is evident.

While the centrality of property relations runs throughout this book, a Marxist analysis does not take us far enough. It is true that property ownership provides economic power. However, the very fact that the Norwegian state introduced the Allodial Law illustrates the way in which different interest groups and political alliances have the power to shape our social existence, independently of economic power. Furthermore, as I will argue in a little while, property ownership provides status and prestige, which in themselves constitute a form of power. It also provides the means for collective organisation and political lobbying, another independent source of power.

Feminist theories offer a useful description of the situation in farming, but offer little insight into how this situation is developed and maintained, and how change occurs. The chapter on the history of dairying uses the example of changes in the dairy industry to illustrate the complex interplay between gender ideology, state policy and access to resources that accounts for the different positions of men and women in the dairy industry. In that chapter specifically, I argue that patriarchy is inadequate to explain the changes that occurred.

## Weber and Power

Like Marx, Weber acknowledges that control of economic resources is one source of power. It provides one possible basis for group formation, collective action and the acquisition of political power. He does not accept, however, that it is the only source of power. Weber argues that power has political and ideological forms as well. Social status or prestige can be, and very often has been, the basis of power. Occupations, political parties, ethnic and religious groups, and styles of life are accorded differing degrees of status, prestige or esteem by members of society. Members of status groups are almost always aware of their common status situation. While class and status constitute two different sources of power, Weber believes that they frequently overlap. In other words, the power that follows from status and prestige is not necessarily based on economic power. He specifically notes that the ownership of property in itself comes to acquire a status value, not in every case, but with remarkable regularity (Weber, 1978). In his study of the sources of social power, Mann (1986) uses and builds on Weber's theory of power. Mann refers to Weber's idea of 'status' power initially as status/ideological power, and subsequently throughout his work as ideological power. I also believe that the source of status or prestige power is ideological. This changes over time, and is subject to crises of legitimacy. When I turn to chronologically review sociological and anthropological studies of land ownership in Ireland, it is clear that the status and prestige attached to land ownership has changed throughout the century. Earlier, it afforded male landowners greater power within their local communities and their families. As we shall see, this has changed. None the less, the status of land holding persists, and landholders continue to constitute a status group as described by Weber. I refer to the work of Hannan and Commins (1992), who describe the far superior social mobility of the children of small landholders compared to the working class. They attribute this to smallholders' self-perception as members of a status group of landowners, and the fact that farming organisations represent their interests, while the working class have no such organisation.

Alongside class and status groups, the third type of group formation that obtains power is that which Weber called 'parties', which he described as living in the house of power. By this he meant that their focus was on political power. Parties, or interest groups, are specifically organised to struggle for power. Parties are specifically concerned with influencing policies and decision-making in the interests of their membership. Parties by definition are organisations, either ephemeral or enduring. They can emerge from and represent the interests of classes or status groups. Equally they may derive from neither. Furthermore, a party may emerge within the structure of the state itself. Within the farming context, farmer's unions represent interest groups organised to struggle for power on behalf of their members. Consisting of property owners, their membership is built on both class and status. The Canadian Farm Women's Network that I will examine in subsequent chapters also represents a party, struggling to influence policies in the interests of its members.

**Parsons and Power**

In one sense, Weber's category of parties represents collective action in the pursuit of power. Power is often narrowly understood as individual power over other individuals, rather than incorporating an analysis of collective power. For all the limitations of the functionalist perspective on power, Parsons has significantly contributed to our understanding of the collective aspect of power, whereby people, through co-operation, can enhance their joint power over third parties or over nature (Parsons, 1960). Parsons argued more generally that inequalities of power are based on shared values. Power is legitimate authority because it is accepted as just and proper by members of society as a whole. It is accepted as such because those in positions of authority use their power to pursue collective goals which derive from society's central values. Hence the use of power may serve the interests of society as a whole. This theory has been criticised on a number of grounds, but most importantly it has been pointed out that power differentials can be divisive rather than integrative, and that it is an arrangement where some gain at the expense of others. The extent to which collective goals based on shared values are pursued has also been questioned.

Let us return for a moment to Parson's useful concept of collective power. Mann (1986) develops two specific senses of power: distributive and collective. Distributive power, or individual power, is where one person has power over another. It is a 'zero-sum' game, with a fixed amount of power, and in order for one person to gain some, another loses some. Collective power is where people, through acting together, increase their joint power (Mann, p. 6). Mann argues that the relationship between the two types of power is dialectical. People in pursuit of goals enter into co-operative collective power relations. In implementing collective goals, social organisation and a division of labour is set up, and distributive or individual power relationships develop. We are unlikely, then, to find a pure case of collective power that does not also include some element of distributive power. We return to this question of individual and collective power when we turn to the question of how power is identified.

## IDENTIFYING AND STUDYING POWER

### The 'Three Faces of Power' debate

What power is, how it is exercised, and how it can be studied have been the subject of an extensive and lively debate in sociology and political science. I believe the debate about the 'three faces of power', which I will spell out over the next few pages, tells us most about how power is exercised and how it is identified. It does not discuss the sources of power in any detail, focusing almost exclusively on its exercise, and how it might be studied. This debate culminates in Lukes' thesis which, as I will contend, is not actually very new, that power can be exercised without visible coercion, grievance or dissent on the part of those over whom it is exercised. Let us quickly review this debate to follow through logically the discussion about how to study power.

The debate we are about to consider began as a critical response to the methodologies, values and findings of a group of scholars who became known as the 'élitists'. Probably the most renowned figure among the élitists was C. Wright Mills, whose study *The Power Elite* (1956) is a landmark in research, with the insights it contains continuing to be influential in the study of politics today. Mills argued that during the Cold War, American society had become dominated by a few individuals exercising their power and influence from a small handful of institutional positions. Mills asserted that the 'power élite' came to share a similar world view and thus to some extent they acted in co-operation with each other. Hunter (1953) applied the élitist model to a study of power and it is this study which illustrates the broad methodological framework for studying élites. People who were believed to be able to recognise the power-holders in the community were identified. They were then asked who they considered to be the most influential people in the community. Those people most frequently mentioned were considered the most powerful.

**The First Face of Power**

Dahl (1958) criticised the élitist model for the imprecise notion of power which it used. It presumed that if an individual was reputedly powerful then he/she actually was so. It presumed the existence of a community. It presumed that power-holders are always visible, and it presumed that power is a commodity possessed by individuals. It also, as succinctly described by Clegg (1989, p. 50), makes power equivalent to the average of some specifically chosen people's perceptions of it. How do we know that what these people think is power, actually is power? If a different group of people had been chosen, would we get the same notion of power?

It was exactly these imprecisions which Dahl sought to counteract. He wished to devise a more precise method of investigating power. Dahl also questioned the idea of 'a ruling élite', saying (quite like Parsons had) that there were different spheres of society which did not necessarily overlap. One could not automatically presume that those who were dominant in, for example, the political sphere, were also dominant in the economic one. Dahl, using his study of New Haven, argued that many individuals and groups were involved in political decision-making, and that participation varied according to the particular issue in question. There appeared to be many centres of power in New Haven. He said the most prominent overall figure was the mayor, and he found little overlap between figures influential in the business sphere and those influential in the political sphere. Dahl claimed that there is a 'pluralist' rather than an 'élitist' distribution of power; it is dispersed amongst many people rather than a few, as the élitists had suggested. Dahl (1961), Polsby (1963) and Wolfinger (1971) were among the main proponents of this 'pluralist' – or as it was later called, 'one-dimensional' – view of power.

Clegg describes the methodology of this approach as being driven by a concern for precision. Precision was to be achieved through a methodological focus on the measurement of power. This model is concerned with the actual exercise of power. This exercise of power is seen to involve observable behaviour, which led the pluralists to study decision making as their central task. They assume that decisions involve 'direct' – that is, actual and observable – conflict. These decisions can then be tabulated as individual 'successes' or 'defeats', depending on whether the individual obtained his/her preference. The participants with the greatest number of successes out of the total number of successes are considered the most influential. This approach was more precise, dealing with visible action, actual behaviour and turning up evidence. There are, however, many limitations to the way it approaches the study of power.

The very emphasis on conflict and decision making (visible action), suggests that power can only be studied in situations of conflict. Dahl (1971) maintains that key political issues should exist, involving actual disagreement in preferences between two or more groups, in order for an analysis of power to take place. If there is no conflict, there is no power play. However, there are many serious problems with this view of power, as initially pointed out by Bachrach and Baratz (1962). They maintain that the undue emphasis on decisions overlooks the fact that power may be, and often is, exercised by confining the scope of decision making. This is visible in grievances rather than conflict. Lukes also criticises this element of the one-dimensional view of power. He says it is highly unsatisfactory to suppose that power is only exercised in situations of conflict, since he maintains the most supreme exercise of power is 'to get another or others to have the desires you want them to have – that is, to secure their compliance by controlling their thoughts and desires' (1974, p. 23).

In other words, what is often paraded as the strongest feature of the one-dimensional view of power, that is, its focus on actual behaviour and conflict, is also one of its greatest limitations. It can only analyse one type of power, and it may be argued that it is even quite a superficial manifestation of power. It is also noticeably ahistorical. It pays no attention to the differences which might already be there in most contexts, prior to a formal exercise of power. The focus on observable action and events means that the pluralists do not examine the historical context of a power situation. They do not explore why different actors occupy different positions, or why they have varying access to resources. This approach assumes that people participate in those areas they care most about and that decision-making arenas are accessible to any organised group (Polsby, 1963).

**The Second Face of Power**

Bachrach and Baratz (1962), however, dispute the approach of the pluralists, saying that access to the decision-making arena is not an open process and that non-participation does not necessarily reflect apathy, lack of concern or consensus. Gaventa (1980) goes further and points out how this notion may result in blaming the victim for his/her non-participation rather than examining the structures which prevent it (p. 8). While Bachrach and Baratz applaud the ideas advanced by the pluralists in the community power debate, they argue that they are too restrictive. They say the study of power must examine both who gets what, when and how, and who gets left out and how (1970, p. 105), and how these are inter-related. While it is important to study decision making, the second face of power necessitates the study of non-decisions. Such non-decisions confine the scope of decision making to relatively 'safe issues' (1970, p. 6). They describe a decision as a choice among alternative modes of action, and a non-decision as 'a means by which demand for change in the existing allocation of benefits and privileges in the community can be suffocated before they are even voiced, or kept covert; or killed before they gain access to the relevant decision-making arena, or, failing all these things, maimed or destroyed in the decision-implementing stage of the policy process' (1970, p. 44). In other words, a non-decision represents an exercise of power because an issue is kept off the agenda, and it is not the subject of a decision, or conflict. This is precisely the exercise of power that the pluralists fail to notice. Schattschneider states the case clearly, saying that all forms of political organisation have a bias in favour of the exploitation of some kinds of conflict and the suppression of others because organisation is the mobilisation of bias. Some issues are organised into politics while others are organised out (1962, p. 949). Bachrach and Baratz elaborate Schattschneider's notion of 'mobilisation of bias', and argue that it is an essential component with which any analysis of power must contend. The dominant values, political myths, rituals and institutions of any community must be scrutinised to reveal whether they tend to favour the vested interests of one or more groups, relative to others.

Michael Mann (1986) has added considerably to our understanding of the mobilisation of bias. He argues that those in power can maintain their position, provided their control is institutionalised in the laws and the norms of the social group in which both operate. Institutionalisation is inevitable, as it is essential to achieve routine collective goals, and hence power also becomes institutionalised. While anyone can refuse to obey, opportunities are lacking to establish alternative machinery to implement their goals. They are 'organisationally outflanked' (1986, p. 7). This brings us back to the distinction between collective and individual power. For Mann, organisational outflanking is the crux of the reason why the masses do not revolt. Because they are embedded in power organisations controlled by others, they lack collective organisation to promote their own interests. In other words, they do not have the knowledge, resources, institutional apparatus or general legitimacy that institutionalised laws and norms provide.

Bachrach and Baratz rejected in advance criticisms that their approach 'to the study of power is likely to prove fruitless because it goes beyond an investigation of what is objectively measurable'. Instead they argue that the pluralists made the mistake of discarding 'unmeasurable elements' as unreal, and by so doing, their approach to and assumptions about power predetermine their findings and conclusions (1962, p. 952). Both Crenson's work (1971), and Gaventa's (1980) study of power in Appalachia illustrate how it is possible to apply the two-dimensional view of power, as it was christened by Lukes, to empirical research. Gaventa illustrates barriers which prevented issues emerging in the political arena, and the resultant non-decisions and non-participation. He further develops this, pointing out that institutional inaction, or the unforeseen sum effect of incremental decisions (the 'rule of anticipated reactions' [p. 15]), influences the extent to which people organise. There is a reluctance to raise particular issues, because the anticipated reaction means protesters view it as a futile exercise.

The second face of power provides a much broader understanding of power than that advanced by the pluralists. Difficulties persist, however, and indeed the two-dimensional model continues to exhibit some of the problems of the pluralists' model. In the same way that the pluralists maintain that power in decision making only shows up where there is conflict, Bachrach and Baratz maintain that non-decision making is visible in the form of overt or covert grievances. Hence the problem of conflict reappears. Lukes asks: 'Is it not the supreme and most insidious exercise of power to prevent people, to whatever degree, from having grievances by shaping their perceptions, cognitions, and preferences in such a way that they accept their role in the existing order of things, either because they can see or imagine no alternative to it, or because they see it as natural and unchangeable, or because they value it as divinely ordained and beneficial?' (1974, p. 24). Thus the most effective exercise of power, that which prevents conflict arising in the first place, will not be uncovered by studying decision making or non-decision making.

Lukes commended the identification of the second face of power, but claimed it did not quite go far enough. There is, he argues, a third face of power.

## The Third Face of Power

Steven Lukes (1974) is the architect of what he has called 'the three-dimensional view of power'. He argues that power can be exercised over an individual by not only getting him/her to do what he/she does not want to do but by influencing, shaping or determining his/her very wants. He says it is crucial for a study of power to consider this dimension of power, for when it is exercised in this way it prevents conflict and decision making from arising.

Initially it may appear impossible to study whether people have wants, thoughts and desires which are different from those they have been shaped into accepting through an exercise of power. However, Lukes argues 'What one may have here is a latent conflict, which consists in a contradiction between the interests of those exercising power and the real interests of those they exclude. These latter may not express or even be conscious of their interests, but, as I shall argue, the identification of these interests ultimately always rests on empirically supportable and refutable hypothesis' (pp. 24–5). The question which must be posed is where such evidence is to be found? Lukes thinks that Antonio Gramsci provides one empirical base from which to make an enquiry when he draws a contrast between 'thought and action, i.e., the co-existence of two conceptions of the world, one affirmed in words and the other displayed in effective action' (Gramsci 1971, p. 326). Gramsci says that where this contrast occurs, it

> cannot but be the expression of profounder contrasts of a social historical order. It signifies that the social group in question may indeed have its own conception of the world, even if only embryonic; a conception which manifests itself in action, but occasionally and in flashes – when, that is, the group is acting as an organic totality. But this same group has, for reasons of submission and intellectual subordination, adopted a conception which is not its own but is borrowed from another group; and it affirms this conception verbally and believes itself to be following it, because this is the conception it follows in 'normal times' – that is, when its conduct is not independent and autonomous, but submissive and subordinate. (1971, p. 24)

Since Gramsci points out that a group, because of submission and intellectual subordination, adopts a concept which is not its own, affirms it verbally and believes itself to be following in 'normal times', Lukes suggests that it would be a valuable exercise to observe how people behave in 'abnormal times', since in an abnormal situation the power apparatus may be relaxed or removed, thus allowing the individual over whom power is exercised to act in a different way. By studying this, we can identify the power structures 'normally' in operation.

Lukes claims that a 'supreme and insidious exercise of power' is that which prevents people having grievances so that they unquestioningly accept their role in the existing order of things. This is achieved because they can see or imagine no alternative to it, or because they see it as natural and unchangeable, or because they value it as divinely ordained and beneficial (p. 24). Lukes further points out that such a bias in the system is maintained by the socially structured and culturally patterned behaviour of groups and practices of institutions. Here Lukes is building on the two-dimensional view of power. While it suggested that barriers exist which prevent issues from emerging into political arenas and thus conflict is restricted, Lukes' approach suggests the bias in the system pre-empts manifest conflict or grievances by shaping conceptions of the situation. The identification of this aspect of power is achieved by more or less following the methods advocated by the two-dimensional view of power, but doing so in greater depth. Predominant belief systems and ideologies must be examined to reveal the less visible ways in which a pluralist system may be biased in favour of certain groups and against others, and also why the latter group do not react. Lukes states this development most clearly in his discussion of Bachrach and Baratz' (1970) study; 'a deeper analysis would also concern itself with all the complex and subtle ways in which the inactivity of leaders and the sheer weight of institutions – political, industrial and educational – served for so long to keep the blacks out of Baltimore politics; and indeed for a long period kept them from even trying to get into it' (1974, p. 38).

Lukes argues that the one- and two-dimensional views of power are restricted because of their presumption that power is possessed only by an individual or group of individuals. He stresses the necessity for studies of power to also consider the power of social forces and institutional practices, since these can operate to create wants, desires and thoughts in individuals. He argues that it is not a question of sociological research 'leading finally' either to the study of 'objective co-ordinates' or to that of 'motivations of conduct of the individual actors'. Such research must clearly examine the complex interrelations between the two, and allow for the obvious fact that individuals act together and upon one another, within groups and organisations, and that the explanation of their behaviour and interaction is unlikely to be reducible merely to their individual motivations (p. 54). Lukes acknowledges that of course such collectivities and organisations are made up of individuals, but none the less the power they exercise cannot be simply conceptualised in terms of individuals' decisions or behaviour. Examples are the control of information through the mass media and through the process of socialisation.

Lukes identifies the difficulties involved in trying to identify one source of power in any given situation. This difficulty arises because there are many factors involved which facilitate an exercise of power, and therefore these factors possess a certain power in themselves. This seems to suggest that any empirical analysis of power relations must aim to identify the various factors which cumulatively lead to an exercise of power, and avoid superficial explanations which would identify one source of power. This is a point reinforced by Gaventa (1980), and more recently by Mann (1986) – it is a meaningless exercise to attempt to identify one source of power in any given situation.

**The 'Three Faces of Power' Reconsidered**

A very rigorous debate followed the publication of Lukes' *Power: A Radical View*.[1] There were prolific contributions for the next ten years, and although the discussion continues, the heat has gone out of the debate. Over the years, even Lukes' most vociferous critics have recognised his work as a valuable contribution to the power debate. The central questions were whether power could be identified when it was people's wishes or interests that were frustrated, how could real interests be identified, and was power being exercised even if people did not resist or complain? The interested reader can follow the debate fully by consulting the works to which I refer in the last footnote. There are only two elements of the debate I will address here, because they directly bear on the study of women on farms. These are real interests, and the question of a situation being accepted as natural and divinely ordained, which basically develops from a belief in the legitimacy of the situation. Both points are in fact related.

**Real Interests**

Lukes' notion of *real interests* is perhaps the most contentious part of his thesis. Critics have argued that it leads to an acceptance of an observer's assessment of the actors' real interests in a given situation.

The empirical basis offered for the study of real interests is also seen as problematic. Lukes advocates the observance of behaviour in abnormal times, when submission is absent or diminished, and the power apparatus is relaxed or removed. Critics ask how such abnormal situations arise, and whether or not the interests expressed in such a situation are necessarily 'real' ones.

The fact that 'real' interests became the most contentious element of Lukes' theory is somewhat ironic, given that it is a concept used in various guises by social scientists for over a century. Lukes' arguments reflect some of Weber's work. Weber argues that domination follows from power, and the most stable power pattern is one where both rulers and the ruled *believe* in the *legitimacy* of domination. In other words, dominant legitimacy rests on belief (Merquior, 1980). The same question regarding real interests could be posed here: was Weber talking about an instance of belief, where the ruled believe in a situation that is not in their real interests? Rousseau's notion of the social contract equally rests on a belief that it is legitimate. Again, is it contrary to some groups' real interests? Marx's development of false consciousness is a clear statement that individuals subscribe to a system that is *not* in their interest. Gramsci argues that hegemony results in the saturation of a society's consciousness, and that it limits the commonsense of most people under its sway. Again, an implicit statement about real interests is contained in this theory. Foucault (1980) talks about the production of truth. Mann's (1986) description of diffused power as a form that spreads in a spontaneous, unconscious, decentred way through a population, resulting in similar social practices that embody power relations but are not explicitly commanded (p. 8), also implies a statement about real interests. In other words, legitimacy, false consciousness and diffused power all embody covert statements about real interests. It is Lukes' attempt to discuss this question openly that has embroiled his theory in controversy.

This is not to deny the genuine dilemmas that a discussion of real interests generate. John Gaventa (1980) presents probably the most novel way around the question of real interests, and it is the one that will be used in this book. He examined the inequalities, and acceptance of these inequalities, in an underdeveloped central Appalachian valley. His work is an example of how the empirical application of a certain theoretical model helps to further refine and hone the theory. Gaventa is careful to stress that the interpreter does not have the right to impose his/her interpretation of what are B's real interests. Gaventa focuses on the *obstruction of choice* rather than trying to identify *real interests* – if it can be shown that B was prevented actively and consciously choosing his/her interests, then Gaventa suggests that it is reasonable to assume that those expressed by B are probably not B's real ones. Throughout this book, we will look at the social customs, particularly the patrilineal line of inheritance, that deny women access to the land. Property confers power, and it could be argued that it is in women's real interests to have as equal an access to the land as men. However it is unnecessary to make this claim. What can be claimed with certainty is that women are unable to make the same choice about becoming farmers, as the system of inheritance prohibits access to the key resource, land. Ironically, the criticisms of Hindess (1982) and West (1986) that Lukes presumes that people from different cultural and social origins would end up with the same constellation of authenticated interests, is useful here to illustrate the third face of power; women, regardless of their interest in farming, their social class or political belief, all face limited access to the land.

**Legitimacy**

Integral to the discussion of the power of social structures and organisations is the power of prevalent belief systems and ideologies which cause people to unquestioningly accept their role in the existing order of things 'because they value it as divinely ordained and beneficial' (Lukes, 1974, p. 24). Lukes acknowledges the importance of legitimation when he quotes Dahl regarding the rule of the patricians in the early nineteenth century – 'the élite seems to have possessed the most indispensable of all characteristics in a dominant group – the sense, shared not only by themselves but by the populace, that their claim to govern was legitimate' (p. 23). While Lukes discusses situations that are considered divinely ordained and natural as those where power can most easily be yielded without any expression of grievance, he does not elaborate the basis of legitimacy in any great detail.

Weber's development of legitimate authority is somewhat controversial. His view that authority always rests on a legitimising belief on the part of subordinates is particularly problematic, as authority may also rest on coercion or force (Weber, 1978; Merquior, 1980). None the less, Weber's assertion that the most stable forms of social relationships are those in which the subjective attitudes of the participating individuals are directed towards the belief in a legitimate order are generally accepted. That the stability of power depends on legitimacy is an argument made by many subsequent sociologists (Stinchcombe, 1968; Giddens, 1968; Habermas, 1973; Mann, 1986).

In the development of his theory of power, Foucault directly discusses legitimation; 'there are manifold relations of power which permeate, characterise and constitute the social body, and these relations of power cannot themselves be established, consolidated nor implemented without the production, accumulation, circulation and functioning of a discourse ... we are subjected to the production of truth through power and we cannot exercise power except through the production of truth' (1980, p. 229). The discourse of truth is for Foucault one of the primary foundation stones of what he calls a 'society of normalisation'. This discourse of truth and society of normalisation leads to the essential credential necessary for the successful operation of power, that is, legitimate authority. The perceived legitimacy of beliefs and social structures greatly inhibits the development of grievance and dissent.

In his study, Gaventa illustrates the link between belief systems and social structure. He indicates that the instillation of an ideology serves to more permanently shroud inequalities and ensure the non-challenging participation of the non-élite (1980, p. 81). Gaventa also recognises the limiting form self-knowledge can take as an indirect mechanism of power's third dimension. He says 'B, the relatively powerless, may recognise grievances against A, the relatively powerful, but desist from challenge because B's concept of self, group, or class may be such as to make actions against A seem inappropriate' (p. 20). It appears then that ideology, belief systems and processes of legitimation restrict the likelihood of grievance and dissent.

When we turn to look at women in farming, the perceived legitimacy of men's hold on resources, access to organisations, the media, and education is quite remarkable. We will consider the manifold sources of this legitimacy, and the beliefs on which it rests.

Most importantly, a discussion of legitimacy leads us to an understanding of how power relations are not static, and how change occurs. Habermas defines legitimacy as a political order's worthiness to be recognised (1976, p. 178). This definition, as he points out, highlights the fact that legitimacy is a contestable validity claim. The stability of an order of domination depends on its *de facto* recognition. We only have to think back through history to the rise and fall of religious power, or the change in the situation of women, to appreciate that legitimacy claims change. I will examine the changing legitimacy of farming structures, with particular reference to Norway and Canada. The former represents a change introduced by the state, where the legitimacy of discriminatory social customs is incongruent with a social-democratic ethos. The latter represents change insti-gated by the organisation of farm women, best understood in terms of a social movement. The key point is that the legitimacy of particular social arrangements change over time. Once the legitimacy of power relations is questioned, it is likely that change will occur.

## WOMEN ON FARMS AND THE STUDY OF POWER

The way in which the above analysis of power relates to the study of farm women can be summarised in four points. Firstly, power means having access to resources which enables someone to control his or her environment. Secondly, power is pursued by organisations and has a distributive and collective dimension. Thirdly, power arrangements are legitimated and they are changed by challenges to their legitimacy. Fourthly, the state upholds men's customary access to land. In any given point in time, power is found in a balance which appears to be consensual. Yet, as Barker and Roberts (1993) so eloquently put it, examining the third face of power allows us to investigate what lies below the concealing draperies of alleged consent.

It is a straightforward task to trace the three faces of power, and conclude that it is the final stage that proves most useful for an analysis of women on farms. The first face concentrates on who makes decisions, and who wins and loses. To focus only on this would mean that we need not consider the absence of women from decision-making structures. A quick glance at farming illustrates that it is a very male occupation, but the focus on decision making does not explain why it is so. The second face of power shifts our attention to what groups and issues are excluded, and what expressions of griev-ance exist. When I look at women's role in farming organisations, I argue that farming organisations are powerful political lobby groups, but that they only include men on farms. The property owner is seen as the producer, and it is these interests that are represented centrally. The absence of women is made visible through the overt organisation of women into alternative associations. This constitutes an 'expression of grievance', in the sense developed by Bachrach and Baratz. The broader agenda of farming issues addressed by these organisations highlights the limits of 'normal' farming concerns, and also serves to illustrate that usually it is only partial farming concerns that are represented. In other words, we do see people, women and particular issues being organised out of farming structures. There is an expression of grievance in some instances, though not all. The third face of power takes the analysis further.

Lukes' theory prompts us to examine the taken-for-granted features of farming. It brings into question the system of land transfer, generally accepted as beyond question, and considered normal and legitimate. It is the source of differing positions in farming organisations, and in education and training programmes. It is possible to identify how patterns of land transfer obstruct the choice available to women. It is less of an option for women to farm in their own right. Patronymic customs are taken for granted, and again they are crucial to the accepted legitimacy of land transfer between men; it is passing land to men that ensures it retains the 'family name'. These are some of the injustices built into farming culture, and they are so taken for granted that there is often no or little expression of grievance. This is the subject of the following chapters.

The discussion of individual, or zero-sum expressions of power, the collective or created expressions of power, as well as 'organisational outflanking', are concepts which are key to understanding the power relations in family farms that will be discussed in the following chapters. When I turn to look at women's relationship to property in twentieth-century Ireland, I outline the zero-sum power relationships in the family and in the community that followed property ownership in the earlier part of the century. I also detail how these have changed. Examples of collective power are evident when I turn to farm-women's organisations. Farm-women's organisations also provide an example of organisational outflanking. Organisational outflanking is central to an understanding of the limited success of the protestations of women in dairying in the late nineteenth century.

To speak of power is to speak of change. The basis of legitimacy of power arrangements is always open to question. We see change occur through a variety of means: through direct intervention by the state, through the changing ethos and commitments of the general society spilling over into the lives of women, and through collective action by women. When I look at women's changed position in dairying almost a century ago, it is clear how this was shaped by their limited access to land and the gender ideology of the time which legitimated their movement out of the industry. A century later, the

Norwegian state's commitment to the social-democratic principle of equality has lead to the revision of the Allodial Law, which transforms gendered access to land. Power relationships are open to change. The following chapters detail change that has occurred, and also identify possibilities for change.

# 3 Property, Power and Women

## INTRODUCTION

In the last chapter, I argued that property is a central source of power in farming. In this chapter I want to turn more specifically to an analysis of property, and in particular look at how a general acceptance of Locke's notion of private property is still current today, and affects women's relationship with land. Locke advanced the protection of private property rights by the state as the protection of a liberty, and he put forward an individualistic, masculine understanding of property ownership. All of these points arise in later chapters.

Property is central to farming. In order to farm, it is necessary to have access to land. However, property is not a commodity equally accessible to all. The majority of people who farm have not acquired their land through the market, but rather through inheritance. There is a closed system of inter-generational transfer of land within families whereby the land is passed on, in most of western Europe and North America, intact to one heir. Farmers, as de Haan (1994) notes, are the socio-economic group which displays the most pronounced occupational heredity. It is extremely difficult to enter farming through the market.

As a result, the transfer of land is governed by a system of beliefs and practices integral to the farm family. The way in which land is transferred, and to whom it is given, is an expression of cultural ideas. Specific cultural meanings ascribed to the relationship between land, farming and the family, shape the reproduction of property relations (de Haan, 1994). Most incredible, as Kennedy (1991) notes, is the resilience of customary practices of land transfer; despite the shifting boundaries between the family, market and state provision. The persistence into the late twentieth century of the kinship-impregnated institution of inheritance in modified though still robust form is nothing short of remarkable (p. 496). It is extraordinary the extent to which traditional patterns of land transfer have been maintained in the face of global or dominant cultural and economic principles.[1]

The most noticeable feature of land transfer is that women rarely inherit land. Customary practices of land transfer distinguish fundamentally on the basis of sex. De Haan asks why inheritance has not been of greater sociological interest, wondering 'how can it be possible for individuals to acquire wealth and status by pure accident of birth, requiring no effort or achievement, in democratic society?' (1994, p. 153). What de Haan does not mention, but which is also intrinsic to the process of inheritance, is pure accident of sex.

There is nothing inevitable about the patrilineal line of inheritance, and indeed Arensberg and Kimball (1940) note that the system adopted in Ireland is only one of many systems that could have been adopted. Nor are systems of land transfer cast in stone. The form land division takes is shaped by the need for survival, as Fox's (1978) study of Tory Island attests. Goody et al. (1976) detail how varying structures of ownership emerged in different places, depending on the economic, legal and political system of a given society. Even in Ireland, the patrilineal system is relatively recent. Prior to the Great Hunger, land was subdivided between children, and impartible inheritance, or the passing on of the farm intact to one heir, was only instituted in the post-famine era (Connell, 1950; Kennedy, 1973).

There are generally more similarities than differences in patterns of land transfer in Europe (Goody et al., 1976). The greatest variation occurs along regional lines, with northern Europe more likely to practice some form of impartible patrilineal inheritance, while southern Europe is more likely to practice partible inheritance (the division of land between children). While partible inheritance seems to be more gender equitable, studies of farming in southern Europe, where this practice is adopted, tend to indicate that gender inequalities of inheritance remain (Oliveira Baptista, 1995; Kasimis and Papadopoulos, 1994). All of the examples and case studies used throughout this book refer to the practice of impartible inheritance. There are different forms of impartible inheritance, with non-inheriting heirs sometimes receiving monetary compensation, or educational provision. There is a great deal of research that remains to be done about differences in inheritance patterns (de Haan, 1994). We know little about the roots of partible and impartible patterns of inheritance. What is clear is the persistence of cultural patterns of land transfer in the face of liberalism and globalisation.

Of course land is not the only form of capital transfer in rural societies: another established system was that of dowry payments. Sometimes the dowry one or more daughters received is presented as a significant inheritance, shaping the life chances of the woman involved. However I tend to agree with Breen (1984) that the inheritance of a dowry was fundamentally different to the inheritance of property. Breen cleverly distinguishes the status and social implications of inheriting movable and immovable capital: daughters were dowered with cash, sons inherited the land; 'Heirs may be seen as occupying a static position in the status hierarchy, while daughters were "floating" – their eventual status would be that of the husband, to whose family they became attached at marriage' (pp. 291–2). De Haan makes a similar point when he notes that the house personifies the link between the living and the dead, between past and future generations, and gives a family an identity in time and space (1994, p. 9). It is men who inherit property and transmit that bond of identity between the family and property.

The protection of traditional cultural practices of land transfer, the preponderance of male heirs, and the protection of private property rights by the state, can all be illuminated by reference to the work of John Locke, and this is the subject to which I now turn. There are four components to this chapter; firstly I want to return to and review the work of John Locke (1632–1704). Locke presents a powerful argument justifying the legitimacy of private property, and presents the protection of this by the state as a liberty rather than an injustice. Locke's notion of private property is premised on the idea of individual property rights. In principle, this individualist notion need not disadvantage women, although in keeping with the time during which he lived, Locke assumed that owners would be men. Yet patterns of land ownership that have fixedly remained in place clearly disadvantage women. Individual women rarely own land. The legacy of Locke has left us with few alternative arrangements to individual private property. Hedley (1982) points out that the term 'family farm' is a misnomer, as ownership of farm capital usually belongs to the senior male in the family. In addition, the relationship between the owner and land is not simply one of subject (owner) and object (land). It is far more complex, and also shapes the relationships between owners and non-owners, which fracture along class and gender lines. To paraphrase Goody (1976), the manner of splitting property is the manner of splitting people, and it has consequences for the position of women, the structure of family roles, strategies of family organisation, and the organisation of economic, political and ideological forces in the wider sphere of social life.

The second, third and fourth components of the chapter are respectively, the culturally accepted social norms, economic factors and the legal framework that shape the transfer of land. The transfer of land is a complex inter-personal process (de Haan, 1994; Salamon, 1987; Commins and Kelleher, 1973; Sheehy, 1980). Property is transferred according to culturally accepted social norms and customary practices, and kinship ideology (Bouquet and de Haan, 1987). Social norms dictate that the heir is male. What I want to look at is the way patterns of inheritance are shaped and negotiated. Economic considerations prompt impartible inheritance. Capitalist intensive farming in northern Europe and North America inspire the keeping of resources together and minimising loss of capital at farm succession and inheritance. According to de Haan (1994), economic pressure and financial problems oblige farm families in these areas to preserve the unity of the farm and reduce claims from non-successor co-heirs. Finally, I will look at the legal system, which fundamentally regulates the transfer of property. Recent changes in Norway serve to illustrate that even where the state appears *laissez-faire*, it none the less tacitly upholds patterns of land transfer. I argue that access to property is the single most important factor shaping the role of women in agriculture. To understand the institutionalised nature of private property, it is necessary to return to John Locke.

## JOHN LOCKE AND PRIVATE PROPERTY

John Locke's thesis on private property has been described as one of the most influential explanations of the origins of private property (Newby et al., 1978; Dunn, 1984). Although the *Two Treatises of Government,* in which Locke advances his ideas on private property, were first published in 1689, many theories of property and political society continue to take his premises for granted (Shrader-Frechette, 1993). One of the problems Locke sets himself is how men (*sic*) can come to have a private right to property. His argument rests upon the 'natural rights' of man: the right to use whatever in the natural environment men deem necessary for their needs, and the right to take as one's own whatever one has worked on, the material objects with which man has mixed his labour:

> though the earth and all inferior creatures be common to all men, yet every man has a 'property' in his own 'person'. This nobody has any right to but himself. The 'labour' of his body and 'work' of his hands, we may say, are properly his. Whatsoever, then, he

removes out of the state that Nature have provided and left it in, he hath mixed his labour with it, and joined to it something that is his own, and makes it his property. (*Two Treatises of Government,* II, paragraph 27)

At least initially then, from Locke's perspective, those who possess more will be those who deserve to do so and they have nothing to apologise for to those who deserve and possess less, as entitlement and possession are tied to the individual's labour.

The right to possess private property is in turn protected by government. Locke develops the notion of government as a trustee for its citizens (O'Connor, 1952). The protection of human entitlements is the purpose of governments. Government exists to safeguard the lives of its trustees, their liberties and their material possessions. The right to private property, then, is upheld and protected by the government. No government has the right to dispossess a man of his rightful property. Following on from this, the government protects patterns of inheritance, as individuals have the right to dispose of their property, once they have made it theirs, as they see fit, and Locke develops the principle that a man's legitimate heirs have a right to inherit his property (*Government*, II, paragraph 138). Locke's theory of private property fuses together the notion of entitlement and merit, and protection by government as the protection of a liberty.

Radically different perspectives on the right to private property and its protection by the state have been advanced. In the nineteenth century, Marx and Engels put forward their very contrary perspective, arguing that private property is the basis of economic and social inequality. The division of society into classes is a consequence of different relations to private productive property by owners and non-owners. Marx argues that private landed property should be abolished, as the root of landed property is sordid self-interest. Marx refutes the notion that individuals who possess land have the right to take possession on the basis of having mixed their labour with nature, arguing that if this were the case, then every human being ought to be a landowner (Marx, 1982 [1844], p. 229). Opposing Locke's contention that state protection of private property is the protection of a liberty, Marx argued that the protection of private property by the state is the protection of privilege and inequality.

Following in a similar vein, Newby et al. analysed stratification and class in rural England, and concluded that property relationships were intrinsic to class relationships. The transfer of property is the transfer of wealth, and of a class position. Furthermore, they argued that Locke's thesis is inadequate to justify the ownership of private property, for if property was that which man has 'mixed his labour with', then how can it be either natural or just for the privileged few to expropriate the surplus created by the labour of the many agricultural labourers (1978, p. 22)?

Over time, other flaws in Locke's thesis have been highlighted. It is criticised for its individualistic notion of possession, particularly with regard to native people and their concept of ownership (Voyce, 1996; Tully, 1994). Native societies, according to Locke, had not developed an advanced political society based on a social contract. Furthermore, native lands were seen as being in a state of nature, and therefore could be appropriated by individuals without the consent of natives. In this way, Locke's concepts of political economy and property were constructed in contrast to Amerindian forms of nationhood and property, and in such a way that they obscured and downgraded the distinctive features of Amerindian polity and property (Tully, 1994, p. 166). The consequence of this, Tully argues, was that the Amerindian system of property in their traditional territory was denied and replaced by a so-called natural system of individual, labour-based property. This provided legitimation for European appropriation of their land without the consent of the First Nations (p. 178). Similarly, the Lockeian idea of individualistic possession was seen as justifying Aboriginal dispossession in Australia (Voyce, 1996). The Australian Crown did not recognise pre-existing native title, and native peoples had to show a proprietary interest in land ownership to have a legal interest in land (Voyce, p. 19).

The sexist assumptions underlying Locke's thesis on private property have also been highlighted. Locke's major premises on ownership of property and inheritance function to preserve a dominant sex, as well as class, position (Clark, 1979). While Locke discusses marriage contracts based on mutual equality in relation to the rearing of children, he says less about mutual equality regarding the distribution of property. Equality in marriage is not compatible with a system of property distribution accomplished through blood lines determined by the father. The exclusive male 'right' to dispose of family property can be established and maintained only by inequality in marriage, at least concerning property (Clark, p. 30). Thus Locke's justification of inequality in the distribution of property is also flawed with regard to the unequal distribution of property between the sexes. Women labour on the land too, yet they do not have rights to the property with which they have mixed their labour. When the state protects private property rights, it protects both class and gender inequalities.

These three critiques of Locke contain the relevant components for an analysis of the relationship between women on farms and farm property. Firstly is the economic or class-based critique. Marx, Engels, and more recently, Newby et al., highlight the economic inequalities embedded in ownership of property. Property provides access to a valued resource, and economic and social relationships are structured around an individual's position in relation to property. Private property owners have economic power over others. This system of inequality is protected by the state.

Secondly, Locke's individualistic notion of property ownership is unable to accommodate alternative ways of owning or managing property, as illustrated by the case of native peoples (Tully, 1994; Voyce, 1996). The same principle is in effect on family farms. Hedley has argued that the term 'family farm' belies the ownership structure that underlies the household. A situation exists on the family farm where there are owners and non-owners of productive property (Hedley, 1982, p. 15). The accepted individualistic idea of property ownership cannot accommodate ideas of family ownership any more than it can accommodate native people's alternative conception of ownership.

Thirdly, the relevance of the feminist critique is self-evident. The thesis of property advanced by Locke is based on sexist assumptions. While Locke's views were obviously coloured by contemporary culture, the continuing currency of his premises lend this critique particular strength. The patrilineal line of inheritance persists, property is owned by men, and women have limited access to this resource. The protection by the state of the institution of private property upholds the disadvantaged position of women in this respect. In addition, Locke's conception of property as a relationship between a subject (owner) and an object (property), involving a bundle of rights regarding use and disposal, has gendered implications. Property is not merely an object, however, it also embodies social relationships between owners, non-owners, and the state (Collins, 1988). The value of property refers not to the inherent quality of an external object, but to the socially and legally defined rights which are attached to such objects. Newby et al., and indeed Marx, elaborate the subject–subject relationships which embody class and economic equality. Voyce (1996) and Tully (1994) argue the subject–subject relationship regarding native peoples. The gendered component of the subject–subject relationship is clearly articulated by Goody (1976); the different arrangements over marital property and the role that these allocate to women in their interaction with men and things have consequences for the position of women, the structure of family roles, the behaviour of kin and the strategies of family organisation (p. 35). This is of particular relevance when we turn to women on farms.

**Property and the Family**

We have noted Locke's very individualistic notion of property. We have also critically noted that he understood property ownership as embodying a relationship between a subject (owner) and an object (property). In fact property ownership affects social relations between owners and non-owners, with particular implications for women on farms. Social patterns of land transfer mean women rarely inherit, or own, land. The interesting sociological question that arises is why women remain part of a social structure that embodies unequal access to property. Women continue to 'marry into' farms, and participate in a social system where land is transferred to their sons. While the increased participation of women in the labour market has somewhat reduced the financial dependence of women on farms, the continued patrilineal line of inheritance persists.

The ownership of property is just one facet of farming. It provides access to the market, and participation in farming organisations. Participation in the public domain of farming is clearly linked to control of the farm. However, farming is generally described as 'family farming'. The term itself is contentious, in that it misrepresents the ownership structure that underlies the household. While this is true, it is also true that the farm household exists on the farm and contains numerous other relationships than that of owners and non-owners of property. Despite the inequalities of the system, women do 'marry into' farms, and participate in a customary practice whereby land is transferred to their sons. Clearly many women marry into farms because it is what they want to do. The power of affection and love towards husbands and children are part of this story. In addition, many women on farms love their work and their outdoor lives (Shortall, 1992). Yet this does not detract from their unequal access to property, and their subsequent limited options to farm in their own right.

Polanyi's work (1945; 1957) provides an understanding of transactions other than those connected to the market. His idea of reciprocity illuminates household and farm household exchange. Polanyi argues that the main form of transaction other than the market is one of reciprocity, where production and distribution of services and materials are undertaken without consideration of profit, but rather long-term give and take. Within the closed group of the family, the principle of organisation is the satisfaction of the wants of the members of the group. This system is not driven by gain, which Polanyi argued was a motive peculiar to production for the market. Relationships between husbands and wives within the farm household, as in households generally, are based on affection and mutual support. There is a common bond of affection for each other, and frequently for the farm. The farm provides for all members of the family that live on it, and while the status and esteem that accompanies land ownership is attached to the owner, it also provides some status for the family in the community.

It was in response to this view of the family as based on reciprocity and complementary roles that many feminist scholars presented the family as fundamental to women's oppression (Barrett, 1980; Young and McCullagh, 1981). Women's oppression was linked to their child-caring role and their financial dependence within the family. It was argued that the ideology of 'the family' blurred women's dependency and subordination. Others have more specifically indicated how women's position within marriage and the family shapes their relationship with property (Engels, 1973; Pateman, 1988; Mulholland, 1996).[2]

The behaviour of farm women exemplifies a struggle between bonds of love, affection and support, and exploitation within the farm family. It is the contradiction between the inequalities that follow from male ownership of property, coupled with the reciprocity that follows on from relationships of affection and love, that partly accounts for the dilemma of many farm-women's groups and organisations that eschew the label 'feminist', even though their actions and aspirations are feminist (Shortall, 1994). The feminist persona of farm-women's groups is complicated by their determined efforts to save the family farm and the farming industry, which in many respects is the source of their unequal position. American farm women, interviewed by Rosenfeld (1985), disliked what they saw as the women's movement's disparagement of the traditional family and of women's role within it. These women saw themselves as part of the *family* enterprise (1985, p. 268). Similarly, the primary concerns of the American farm-women's group WIFE are the family farm and farm family welfare rather than themselves as farm women (Haney and Miller, 1991). Australian farm women have similar problems with the idea of feminism (Alston, 1990; James, 1982). Hostility towards men is inconceivable in a system that depends on such a high degree of co-operation. Furthermore, farm women and their husbands are joined by marital ties and bonds of affection and do not want to become business adversaries. Feminism is very unpopular in the Australian bush because it seems to contradict the relationship between farm women and their families (Alston, p. 24). Editorials and letters' pages in the newsletters of the Canadian Farm Women's Network's provincial groups also reflect women struggling with their commitment to the family farm *and* their pursuit of individualistic feminist objectives, such as dismantling the barriers to women's participation in agricultural organisations (Shortall, 1994).

One of the shortcomings of Polanyi's concept of reciprocity is its lack of any perspective on power. While at one level reciprocity occurs within the farm family household, it exists alongside a very unequal relationship of economic and social power. Whatever the bonds of affection and love, and satisfaction that women on farms derive from their work, it remains the case that there is a fundamental inequality within the structure of farming and the farm family. Farming is an occupation that denies women access to property, its central resource, and consequently to the market and economic gain.

Partly as a result of familial reciprocity, the gendered inequality that exists in farming is taken for granted. This resonates with Newby et al.'s argument that ideologies of property ownership contribute to a system of 'natural' inequality in the countryside which can remain an extraordinarily prevalent feature of the taken-for-granted perception of rural society (1978, p. 25).

## ACCESS TO THE LAND: CULTURAL NORMS, ECONOMIC ENVIRONMENT AND LEGAL CONSTRAINTS

### Cultural Norms

Locke's notion of property advances the idea of an individual property owner. He also applauds state protection of private property and the transfer of private property to heirs. This process of passing on land is regulated by customary cultural practices. Research has illustrated the complex inter-personal nature of land transfer (Commins and Kelleher, 1973; de Haan, 1994; Sheehy, 1980; Salamon, 1987). Property is transferred according to culturally accepted social norms and customary practices, and kinship ideology (Bouquet and de Haan, 1987). Culturally accepted social norms and customary practices dictate that the heir is male. The most significant feature of inheritance practice is that property is transferred to men. It is very rare that a woman with brothers will inherit land (Shortall, 1992; Kennedy, 1991). A key element of the accepted cultural norms surrounding the transfer of property is that it is passed from father to son. The main route of entry to farming for women is through marriage. The majority of the small number of women who own land in their own right are widows.

Farmers have strong ties to the land, and many view it with sentimentality (Salamon, 1987; Commins and Kelleher, 1973). For many, 'keeping the name on the land' is an important goal, and it is only through passing land to sons that this desire can be fulfilled. Research throughout this century has illustrated the accepted legitimacy of the transfer of property to sons. Central to the understanding of farming as a male occupation is access to property. Kinship systems of land transfer play a key role in constructing men and women differently in this regard (Whatmore, 1991), with sons, regardless of interest or ability, seen as the natural heirs. Sons are prepared for succession, and included and trained in farming activities. As I have commented earlier, ownership of property is not only about the relationship between the owner (subject), and property (object). It shapes social relationships between family members, and relationships in the wider community. Interestingly, most studies of farming, and in the next chapter we will see this in the case of Ireland, have focused on the relationship between fathers and sons in relation to property ownership. These studies examine the negotiation of control, status and prestige that accompany the transfer of property. The gendered nature of land transfer and ownership is less of a concern, even though women's relationship to property fundamentally shapes their farming lives. Within any family farm, the cultural norms of land transfer mean that the owners and non-owners, and the potential owners and non-owners, are fractured on the basis of gender.

Access to property is not without costs. Often the son, and this was particularly true in the past, has to wait a long time for the father to hand over property (Arensberg and Kimball, 1940; de Haan, 1994; Breen et al., 1990). This had implications for the heir and also the structure of agriculture more generally (Commins and Kelleher, 1973). Inheritance also brought a moral obligation to stay on the land and care for elderly parents, thus restricting other employment opportunities, and in the past, marriage opportunities. Brody's (1973) representation of inheritance is novel in that he not only presents ownership of property as a source of power and privilege, he also presents it as a responsibility which obstructs the life choices an heir can make. Inheritance entails a responsibility to continue the family farm and remain in the traditional society which, he argued, was becoming increasingly demoralised. While this situation is no doubt true for some heirs, particularly on smaller farms, there is little reason to believe it is the norm. The cultural adaptiveness of the small farmer community has been much more in evidence than the cultural and social–psychological demoralisation described by Brody (Hannan and Commins, 1992). Whether the transfer of property is understood as a negative or positive process, it remains fundamentally gendered.

The tenacity of the system of land transfer not only relates to the fact that property provides the economic resource necessary to farm. Historical and contemporary studies of farming attest to the status and prestige of land ownership, and the security it symbolises. It is this, along with the economic significance of land ownership, that has led to tenacious patterns of land transfer. Owning land rather than its productivity is the key source of status and prestige. In looking at why many Irish smallholders with commercially non-viable holdings have held on to their land, Hannan and Commins conclude it is because the symbolic significance of being a landholder and its status implications probably count for more than its purely economic returns (1992, p. 84). This has added to the high level of attachment to the land, and has directly shaped the structure and development of agriculture, as only a limited amount of land comes on the market (Hannan and Commins, 1992; de Haan, 1994). Kennedy argues that the institution of inheritance has acted as a breakwater against market pressures and the ultimate triumph of agri-capitalism (1991, p. 498). For many years, farmers were reluctant to retire, and low-income small farmers do not exit from farming, but retain their farms for reasons of status and prestige, and its symbolic importance. Studies of inheritance patterns highlighted the important implications for agriculture generally of this system of late land transfer (Commins and Kelleher, 1973). States have tried to influence these patterns, through early retirement schemes for farmers, and in Ireland, through the Land Commission which endeavoured to provide an alternative means of accessing land

and structuring land holdings. However, policies relating to land restructuring have been restricted by the rights of property holders, the high level of attachment to land ownership, and the importance accorded to the possession of land as a source of status and prestige.

The way in which landowners are a status group in Weber's sense is clear from Hannan and Commin's study of how small Irish land-holders have fared during the socio-economic transformations from the 1960s on. Compared to the working class, small scale landowners were in a better position to use the new opportunities that became available. Small landholders retain a strong, self-confident sense of their own culturally independent importance and status within Irish society (1992, p. 97). Their identity is shaped by being landowners, regardless of size of the holding. Farmers' organisations have always maintained easy accessibility to neo-corporatist arrangements for the negotiation of state policies, and while the issues articulated may primarily reflect the interests of the larger farmers, the concerns of smaller farmers none the less receive greater attention than those of the working class (p. 101). Small landholders have benefited dispro-portionately from industrial expansion policies, and the widespread provision of education. Small landholders have consistently won the competition with the working class for industrial jobs, as well as for upward education and social mobility for their children (p. 103). In other words, their self-identity as a status group has led to higher mobility aspirations, and the perception of less constraining social and geographical barriers to their attainment, than among the working class.

**Economic Considerations**

Inheritance is not only shaped by cultural norms. Economic consider-ations also influence patterns of inheritance. The transfer of land is driven by the desire to ensure the survival and continuity of the family farm (Gasson and Errington, 1993). Passing on the farm intact to one heir is the most satisfactory way of ensuring this continuation (Kennedy, 1991). As I have discussed earlier, there are examples of different patterns of transfer in alternative economic and legal condi-tions. There are inconclusive debates as to the extent to which patterns of land transfer are shaped by economic considerations. However, it is impossible to ignore the power of cultural practices (de Haan, 1994). Nor are systems of land transfer set in stone. For example, the Irish system of inheritance is relatively recent, dating back to around the time of the Great Hunger, prior to which partible inheritance, that is, the subdivision of holdings, existed.[3] Around 1815, a decline in cereal prices and an improved British market for livestock made it more profitable to put land out for grazing, and the land available for the potato–cereal mix became insufficient to support an expanding population. It was these changed economic conditions that prompted the move from partible to impartible inheritance patterns (Breen et al., 1990). Robin Fox (1978) has described the very different system of inheritance on Tory Island (off the northwest coast of Ireland), tied to economic conditions, where every child of a landowner has a right to a portion of the land. There is no absolute right to property, instead many members of a family retain a claim to the land. The key principle deciding amongst the claimants was consideration of 'land of the marriage'. That is, each marriage required land, and the transfer of land was driven by the attempt to ensure each marriage had enough land on which to survive. Fox described this system as a sensible human adaptation to a difficult environment, in that it allowed for the rational distribution of land which, given the island's population and terrain, would not have been facilitated by the patrilineal system of inheritance. This presents an example of economic factors shaping inheritance patterns. Farmers are economically rational actors, and seek to maintain their existence and maximise their profits. The way in which they do this bears the imprint of their cultural values. In other words, while economic considerations are important, they are filtered through cultural and social norms. Ironically, although economic factors may shape inheritance patterns, there is no logical reason why they should influence the sex of the heir.

## Law and the State

In addition to economic considerations, inheritance is also shaped by legal considerations. In Ireland, impartible inheritance, or the passing on of the whole farm to the chosen heir, is legally intact. Legal regulations vary in different countries. For example, in Sweden, the heir must buy out the siblings' share, and in Norway, the eldest child, irrespective of sex, is the legal heir.

The transfer of land is regulated by the state through the legal system. While the law is established by a political authority and backed by state power, it is none the less generally sanctioned by the society in which it operates, and is reflective of normative behaviour.

In other words, while the legal system regulates the transfer of property, it generally reflects the dominant beliefs and values of the society in which it operates.

While Locke presents the protection of private property by the state as the protection of a liberty, from the perspective of women in farming it is clearly the protection of an injustice. The whole system of individual property ownership requires differentiation between the heir and the non-heirs, and from the position of the many non-heirs, it is unjust. The choice of heir, however, is not made from one amongst equals. There is an inherent gender inequality in the process of selection. The patrilineal line of inheritance is a social custom, yet it is upheld by the state. It might be argued that it is difficult to see how the state could intervene to alter what is essentially a transfer of property deeply embedded in a private social custom within the family. Voyce (1996) points out, however, that the supposed existence of a neutral state and the supposed non-intervention in the private sphere is a misconstruction. In reality, the state has always intervened in the private sphere when so required. Its supposed non-interference in the private sphere leaves existing social hierarchies in place. The power which the state has to influence the social practice of the patrilineal line of inheritance is clearly illustrated by Norwegian law. In 1974 the Allodial Law was introduced in Norway, whereby the eldest child, regardless of sex, is the legal heir to the farm. Compared to this, the way in which the neutrality of other states maintains the *status quo* becomes more visible. I will examine the state again in the penultimate chapter.

## CONCLUSION

Women's relationship to private property, in this instance farm land, is central to our understanding of their position in farming. The legitimacy of private property ownership and our general understanding of private property was profoundly shaped by John Locke. He presented an essentially individualist view of property ownership which continues to prevail, and even though a family lives and works on a family farm, we accept that ownership will be vested in one person. In effect this means that the economic, social and political power that accompanies property ownership belongs to one individual. In addition, that individual is generally male. Whether or not the owner is male or female does not necessarily shatter the basis of

Locke's notion of private property. None the less, social customs have held patterns of male ownership firmly in place to the present day.

The individualistic nature of property ownership transfers over into the popular understanding of farming as an occupation carried out by one individual within the family. This is examined in later chapters where I look at individual men's participation in farming organisations and individual men's participation in agricultural education and training. The focus is on the property owner within the farm family, and the work and participation in the farm by women receives less attention.

Locke presents property ownership as a male right, and it is the right of the owner to dispose of his property as he pleases. This right is protected by the state, which Locke presents as the protection of a liberty. In the instance of the transmission of land, from the perspective of women, it must be understood as the protection of an injustice. There is nothing 'natural' about a system of land transfer between father and son. On the contrary, it is a clear example of a discriminatory social practice that fundamentally distinguishes between men and women. In Chapter 8, I consider how the Norwegian state has directly intervened to change customary practices of land transfer, because they contravened the social-democratic commitment to equality.

The transfer of property is further mediated by cultural norms and behaviour, economic considerations and legal constraints, and the interaction of all three. The power, status and prestige that is attached to land ownership means there is a tenacious hold on the land and the traditional patterns of transfer to sons. While economic factors shape decisions about land transfer, these are mediated by cultural norms, in particular, the belief that land should pass to men. This is the subject of the next chapter. The state, often tacitly, is the backdrop that facilitates particular forms of land transfer. For example, there are instances of the state's trying to influence the amount of land that comes on the market, either through early retirement schemes or methods to re-allocate land. Rarely, however, has it interfered with private property rights through the procedures it has adopted.

Patterns of land transfer and the preferred sex of the heir are not cast in stone. Our social practices and customs do change. Their acceptance at any time rests on a belief in their legitimacy (Habermas, 1976). Changes in the perceptions and status of farming have occurred and may continue to occur, with particular implications for inheritance patterns. Research has illustrated the difficulty on some small farms of securing an heir (Scully, 1971), either because of the more attractive opportunities offered by waged employment, or because farmers have not married. More recent research (O'Hara, 1994) suggests that women on farms are consciously orchestrating the withdrawal of their daughters and future farm wives from farming (p. 64), and in this way, inheritance and succession patterns are threatened, as indeed is the continuity of family farming as a social form.

While the above situations – lack of heirs, lack of interest of heirs, or the lack of women to marry heirs – seriously threaten the continuation of the family farm, I would argue that they do not necessarily question the legitimacy of the social norms and practices that deny women access to property. They threaten the future of the farm, but not the traditional pattern of intergenerational transfer of farms to sons, where sons exist and are willing. The changing legal situation, whereby an increasing, albeit small, percentage of women are becoming legal partners in the farm, may also represent an avenue of change. Research suggests, however, that this changing legal status may simply represent a means of tax avoidance (Whatmore, 1991; Haley, 1988) rather than a deep-rooted change in their position on the farm, and that many women feel their partnership status is nominal (Shortall, 1996). Nor does this partnership alter the limited opportunities for women to farm in their own right. A more overt, and interesting example of change in inheritance patterns is provided by Norway. The Allodial Law, introduced in 1974, decreed the eldest child to be the legal heir to the farm, regardless of sex. The legitimacy of the discriminatory customary practice of property transfer was not congruent with the ethos of the Norwegian social-democratic state, and its commitment to gender equality and fairness (Haugen, 1994). The Norwegian action does not remove the inequalities of private property ownership. What it does do is remove the gender inequalities of property ownership. This action by the Norwegian state also serves to illustrate that the transfer of property is fundamentally regulated by the state, even if that regulation appears to be *laissez-faire*, allowing social norms and patterns to persist unfettered. The protection of private property by the state can be understood as the protection of a liberty, following John Locke, or the protection of a key source of social inequality and prestige, following Marx and Engels. In the case of the acquisition of land, it must be understood as the protection of an injustice, which upholds social norms and traditions that clearly disadvantage women.

In the next chapter, I will consider the social norms and traditions that have governed the transfer of land in Ireland throughout the twentieth century.

# 4 Women and Property in Twentieth-Century Ireland

## INTRODUCTION

I now turn to look at women's relationship to property in Ireland throughout the twentieth century. Undertaking this analysis serves to illustrate the embeddedness of the institution of patrilineal land transfer. It is important to grasp how deep-rooted this tradition and social custom is, as it fundamentally shapes women's role in farming. My case study in this chapter is Ireland. It would be impossible to review all Irish ethnographies, and I have selected some of the better known and most cited studies. However, the issues raised are easily generalised. For example, de Haan (1994) raises similar questions using the Netherlands as a case study, Salamon (1992) does so for ethnic communities in the States, and Goody et al. (1976) do so for western Europe generally.

This chapter looks specifically at property transfer, the culture, norms and values that shape property transfer, and the power property ownership bestows. Taking an historical perspective illustrates the tenacity of patterns of land transfer between men. It also illustrates the strength of the customs that regulate the transfer, and the economic power, status and political power associated with property ownership.

As has already been noted, land ownership is crucial to our understanding of the position of women in farming. Beginning with Arensberg and Kimball's seminal work in the 1930s, anthropological and sociological studies have illustrated the power, status and prestige attached to land ownership. However, the Irish society of today is very different to that described by Arensberg and Kimball (1940), and ownership of land no longer provides the same power within the family, or access to the patriarchal society they described. We see changes in the power associated with land ownership as wider societal changes occur. Access to education and increased participation in the labour force have significantly influenced women's economic and social independence. Yet, while women now have many employment opportunities, most women on family farms continue to enter through

46

marriage. Their position on the farm is still underpinned by their relationship to property. A great deal of research has illustrated that their work is unaccounted for in statistics (Sachs, 1983; Rosenfeld, 1985; Reimer, 1986; Whatmore, 1991; Shortall, 1992; Gasson, 1992; O'Hara, 1994), that they are under-represented in farming organisations (Teather, 1996), and that they are not actively engaged in agricultural training organisations (van der Burg, 1994; Shortall, 1996). The property owner continues to be the recognised farmer, both in customary practice and in dealings with official organisations. It is this that largely renders women invisible.

Throughout this chapter, then, I will examine the status and prestige that accompany land ownership, and illustrate the tenacity of land transfer patterns from father to son. Most of the better known Irish anthropological and sociological works dealing with farming this century will be reviewed. The status and prestige that goes with land ownership is obvious. What is equally clear is that change has occurred over time, and property ownership no longer grants the same power position to the property owner within the community or the family. On the other hand, the pattern of patrilineal inheritance remains as strong as ever, and because of this, women continue to have unequal access to land.

All of these studies attest to the central role of property relationships in structuring access to the public domain of farming, and the economic and social power this bestows. However few of them focus on the gendered structuring of property relationships. Nevertheless, the analyses are sufficiently rich to enable extrapolation. Let us begin with the seminal work of Arensberg and Kimball.

## WOMEN AND PROPERTY IN TWENTIETH-CENTURY IRELAND

### Breaking New Ground: Arensberg and Kimball

Arensberg and Kimball are credited with the provision of the first anthropological account of the main social and economic conditions of rural Ireland in *Family and Community in Ireland* (1940). Prior to this, accounts had been presented only through literary or political commentary and controversy. Their documentation of the social and economic conditions of Co. Clare in 1932, which they considered representative of Ireland, marks the initiation of rural Irish sociological research. Arensberg and Kimball's work sparked considerable controversy, both for its claims that it was representative of all of Ireland, and for their rigid underlying conceptual model of structural functionalism. As they explicitly indicate in their introduction, their work aimed to be both ethnographic, and to advance the explanatory power of structural functional analysis (Arensberg and Kimball, xxv–xxvii; Hannan, 1982). Hannan has conclusively argued that, despite theoretical shortcomings, Arensberg and Kimball's work is an important ethnographic account of the social and cultural system characteristic of small-scale farming communities in the west of Ireland in the 1930s. The focus here is on the insights they give us into property ownership, and the economic and social relationships of which it is the foundation.

Arensberg and Kimball describe the existing patrilineal system and acknowledge that it is one of many possible systems that could have been chosen. The Irish family is patrilocal and patronymic, and farm, house and most of the household goods descend from father to son with the patronym. The father is dominant within the family. He comes to stand for the group which he heads; the farm is known by his name, and the wife and children bear his name (p. 80). Arensberg and Kimball frequently use 'father, husband and farm-owner' as synonyms (pp. 46, 47). They detail the father's 'controlling role' (p. 46), and attribute it to his status as landowner; 'The old man abdicates his controlling position with his transference of the farm to his son' (p. 121). They recount the social standing and precedence accorded to the old fellows, 'the men of full status who head farms and farm-working corporations of sons, those who have turned or are about to turn over their control to a younger generation' (pp. 170–4).

Arensberg and Kimball give us a detailed description of the work carried out by women, which suggests that such work is arduous and time-consuming: 'The first duty of the day falls to the woman. She rakes up the fire and gets it going, and starts getting the breakfast ready' (p. 35). At about 5 o'clock, the work of the men is over for the day but that of the woman goes on (p. 39). She must prepare, serve and clean up after the tea, milk the cows, help the children with school lessons if necessary, and put them to bed. If she returns to join the men at the fire, she continues with knitting and baking. When the whole family has gone to bed she closes up the house and slakes the fire in the hearth. They say too that 'the woman's hands are never idle' and that the work of women is as important in farm economy as men's work (p. 63).

From the perspective of their structural functionalist framework, it is hardly surprising that Arensberg and Kimball present the gender-related division of labour they observed as a functional development within the society. The 'duties of male and female are complementary' (p. 195), and the division of labour between the sexes simply represents the separation of human activity into male and female spheres. They describe the division of labour between the sexes as one that arises within a field of larger interests and obligations. It is 'part of the behaviour expected reciprocally of husband and wife. It is a functional element of their relationship within the family' (p. 48). Within the family the father has the controlling role, and he directs the activity of the family. However, he has an obligation to act in the interests of his wife and children, and should he fail to do so, they are rightly entitled to feel anger towards him.

Despite this account of reciprocity and complementary roles, Arensberg and Kimball recognise the different status and prestige vested in each role. They speak of the farmwife and mother 'who serves her men' (p. 35), who, as they eat, stands ready to refill their plates (p. 37) and who does not sit down to eat until the men have finished. They say, too, that until recently and even still in some more remote districts at the time of their study, a peasant woman kept several paces behind her man. Arensberg and Kimball identify the different status attached to the work sphere as the basis of the unequal access to wider social structures, with the work sphere of men described as more various and of increased value. However, it seems more evident that in the same way dominance in the family is tied to ownership of land, so too is access to wider social structures; it is by virtue of their previous control of the farm, their position as farm heads and owners that the old fellows, the men of full status, come to 'represent the interests of the community before priest, schoolmaster, merchant, cattleman and government official' (pp.170–4). Arensberg and Kimball's functionalist framework causes them to focus on complementary work roles. They do not question the basis of the different valuation of gendered work, which seems clearly to be linked to the ownership of land. They describe the dominant position of the father within the family, and alongside this, provide an anomalous account of reciprocity within the family. This is a charge levied against Arensberg and Kimball by their critics; their concern to assert the importance of structural functionalism means that they provided many descriptions of observed relationships that do not conform with their theoretical model (Hannan, 1972). The questions of property, power, women and complementary work roles are clear examples.

Given that Arensberg and Kimball do not develop the links between property ownership, gender and power, it is interesting that they do so in the context of father–son relationships. It displays their awareness, albeit at times underdeveloped, of the relationship between property and economic and social relationships in the family and the community. They ponder the relationship of the father and son in great detail. They describe the 'lifelong subordination' (p. 56) implied by the retention of the name 'boy' by sons until they are in their fifties. The father has full discretion over which son he will choose to stay on the farm (p. 65), and until the farm is made over to this son, he remains a 'boy'. While sons may go to the fair or market with their fathers, it is the 'farmer–father' who does the bargaining (p. 55). They say, too, that the father/son division of labour is more than an arrangement of farm management: 'It is very directly part of the systems of controls, duties and sentiments which make up the whole family life. The apprenticeship and the long subordination of the sons in farm work are reflections of the entirety of their relationship to parents; it is impossible to treat the two spheres of behaviour separately' (p. 56). This valuable analysis of the position occupied by the son in relation to his father and the way in which it relates to other interpersonal relationships moves far beyond the confines of their functionalist framework. Yet it begs for a similar analysis of the power relationships between husband and wife, and how they relate to property ownership.

While it is never the specific focus of their attention, Arensberg and Kimball display a remarkable understanding of the manner in which women are excluded from property ownership by accepted belief systems. The patrilineal structure is temporarily altered in the case of widowhood, and where there is no male heir. Here the incoming son-in-law must provide a larger fortune than would be demanded of a daughter-in-law since 'the additional payment is deemed necessary to overcome the anomaly felt at the reversal of the usual roles of sons and daughters' (pp. 109–10). They detail, without dwelling on it, the continual disinheritance of women. They note too that the strength of the patrilineal line reveals itself in the aberrant cases where women do control land (p. 125). They say that 'the country districts recognise only vaguely the right of a woman to hold property. The patrilineal identification of family and land is incompatible with it. Whatever farm a woman works or controls is regarded as a trust for a son or brother of her husband or father. The conflict is naturally never so phrased, nor is it ever debated, yet there are many instances which show behaviour, accepted fully as right and just by the local community, which reveal the patrilineal identification' (p. 133). Arensberg and Kimball recognise here that conflict does not have to be stated or debated to exist and they also recognise that it may manifest itself in unquestioned and legitimatcd actions and conduct. This is what is happening to women – it is not considered 'right' for a woman to be a farmer or hold property. The shortcoming of Arensberg and Kimball's analysis is the failure to consider the gender implications of this restricted access to property, or why such sanctions are necessary if the male/female spheres are truly divided in the complementary, idyllic way they suggest.

In conclusion, the rigid functionalist framework adhered to by Arensberg and Kimball meant they present the 'complementary roles' of men and women as having evolved by some kind of natural process. They describe the superior status, prestige and financial reward of each facet of the male role, and how social customs and taboos keep this position intact. Yet it is clear from their study that the different economic, social and political power positions of men and women arise from their different relationship to property. Unfortunately their functionalist analysis meant this was not the focus of their study.

**The Limerick Rural Survey**

The Limerick Rural Survey 1958–1964 was the product of an agreement between the Irish government and the US government which made a limited sum of money available towards the cost of technical assistance projects developed by rural and farming organisations. The proposed Limerick Rural Survey was awarded financial assistance. This survey is a detailed, multi-disciplinary study of Limerick, and the lengthy chapters deal with physical geography and geology, social history, demography, social structure, social provision and rural centrality, information on population and social provision. The overall aims of the survey were to provide a description and analysis of the developmental infrastructure of an Irish county, to provide a study which highlighted the need for similar studies throughout the rest of Ireland, and finally, to demonstrate the importance of this kind of study in playing a vital part in a more rational development of resources. The part of the Limerick Rural Survey which is of particular interest for this chapter is Part IV, 'Social Structure', by Patrick McNabb (1964, pp.193–248). His main aim in this study was to ascertain the social causes of rural migration. He looked at what he considered to be a society in transition, with education providing upward mobility for farm labourers. McNabb's study details how the economic and social life of rural Limerick was structured by property relationships. In particular, he illustrates how class is organised around property. While he pays little direct attention to property and gender relationships, his study is rich enough in detail to allow easy extrapolation.

The hypothesis with which McNabb begins his study is that 'The chief institutions which give the community its specific form and function are so organised and related to each other as to guarantee the authority of the farmer and to maintain the social prestige and traditional functions of the farm family' (p. 194). If this is true, he maintains it should be possible to show that migration and other community problems are related to and affected by the role and function of the farmer and his family. McNabb clearly describes the authority, social prestige and status which are embedded in the role of the farmer/father figure. He says that a man cannot achieve full economic and social status until he becomes the owner of a farm. He points out the source of the father's authority as being his status as owner of the farm, a position which he values dearly. McNabb says that the father is 'loath' to discuss matters of finance and farm management with his wife or adult children, indeed 'their advice or interference would be resented' (p. 228). Elsewhere he says that the father's regard for his authority is so strong 'that he will risk losing his whole family rather than make concessions to their point of view ... his main interest, while providing for the family, is to prevent them from invading his own personal domain' (p. 229). He argues that marriage as a social institution is not simply a contract between individuals. It is also a property transaction which involves the rights of every member of the family, and particularly the father's authority and right of ownership (p. 243). McNabb rightly identifies the source of the father's authority as his position as owner of the farm. He also notes that the size of farm determines position on the social scale. If a farmer wishes his family to maintain its status, he must avoid dividing land amongst his children (p. 225).

He says later that traditional society is well-established 'because it controls the means of production' (p. 244). Changes regarding the increased availability of education actually serve to maintain traditional norms regarding property and the role of the father, because it becomes more legitimate to have one heir, and educate the other children (p. 244). McNabb details the patrilineal line, and does note the exclusion of women from the process of land transfer. He describes in detail the privileges land ownership bestows on the male 'farmer'. He clearly outlines the total exclusion of women from the process of land transference. There is a pattern to the patrilineal process. It is not necessarily decided on the basis of primogeniture: for example, it is 'unlikely that the eldest son will inherit if he shows scholastic ability. The majority thought that the son who had no interest in school would inherit or, if there were a few such sons in one family, then the father's favourite would be chosen' (p. 225). He goes on to say that if the children are young and the father dies, the farm is left to the mother 'to do the best she can' (p. 225). If there is an adult son, he will get the farm, but provisions are made which allow the mother to stay on the farm 'for her day'. The father greatly values his authority and social prestige and holds on to these tenaciously. He is inspired to hold on to them in this tenacious fashion by the fact that he himself has been a dependant for so long. He describes how the father demands absolute obedience from the son, and the son is intensely aware of his subordinate position (p. 231). Thus, 'when he inherited, he got those things which he had been deprived of for so long; owner-ship and the right to dispose of his own personal property' (p. 229) and this results in his determination to prevent his family 'from invading his own personal domain' (p. 229).

McNabb is sensitive, as we can see, to the father's attitude to prop-erty, and how the process of acquisition shapes the son's world view. Yet, relations of gender and property are not so overtly articulated. He details how important the authority position is and becomes to men, but women will never be involved in this process, they will never hold a status position, power position or authority position. McNabb tells us that a man achieves full economic and social status only when he becomes the owner of a farm (p. 243). Yet he does not consider the corollary of this for women. They are virtually assured that they will never become the owner of a farm, therefore they can never occupy a full economic and social status position. McNabb does not investigate the implications of this for interaction, or how it will affect the self-perception of the woman or the way in which spouses will react to each other in the knowledge of these processes.

Although it is clear throughout the study that women are not involved in farming organisations and structures, they do seem to carry out a substantial amount of farm work. McNabb says that the daughter is never compensated for her work on the farm (p. 188). He describes the way in which wives and children take on the farming duties if the farmer is going to town or the market or to a sporting event (p. 233). A wife, however, cannot leave her duties so easily: 'A woman with young children rarely sees the outside of her home' (p. 234). He says that while the husband's working day is broken by visits to the pub, the market, town, hurling matches or race meets, the wife's life by contrast 'is one of unrelieved monotony' (p. 234). Her household duties are 'onerous and unvarying' and added to these are 'the farmyard duties which she must shoulder when her husband is absent. Holidays are unknown. Even if she could take the time off during the day, the community provides no suitable form of leisure ...' (p. 234). It is obvious that the work of the farm wife is time-consuming and restrictive, but also that she is perfectly capable of taking control of the farm when it suits her husband for her to assume this position. It becomes clear that it is not the lack of ability or the shirking of farm work which prevents the farm wife's involvement in farming organisations and structures, it is the lack of recognition she receives for her work. It is the work of the property owner or farmer that is publicly recognised and accorded status.

McNabb's analysis of participation in Muintir na Tire, an organisation for community development which aims at restoring a spirit of co-operation and self-help, displays the clear relationship between economic power or class, and property. He describes how the participation of the farmers and the farm labourers in Muintir na Tire reflects their more general positions in the social structure and relations to each other. The farm labourers complain that the farmers hold centre stage: 'They wanted nobody to have a say but themselves' (p. 208), and keep the front seats for themselves. Many farm labourers left Muintir na Tire because of this. The farmers, on the other hand, say that it is the labourers' own fault: 'They won't speak up when they are at a meeting. They do all their talking outside the door' (p. 208). It is the ordering of seats which prompts McNabb to say that it shows 'how the traditional class structure is revealed in ordinary day-to-day pursuits' (p. 208). It also displays a particular expression of power in action; the stronger group (the farmers) monopolise the platform, and the structure of the organisation inhibits the involvement of the labourers discreetly and indirectly. The stronger group, however, deny this organisation of the situation and instead blame the weaker group; 'They won't speak up' ... 'do all their talking outside the door'. The basis of the power of the two groups is their relationship to property.

McNabb also provides an interesting perspective on the relationship between the state and the owners of property. He says that the organised forces of change in the rural community are not, as might be expected, the various government agencies for the improvement of agriculture. These 'being paternalistic, are part of the traditional framework ... they do not change it' (p. 245). He declares that 'the chief institutions ... are so organised and related to each other as to guarantee the authority of the father and the conservation of property' (p. 243).

In conclusion, McNabb says that the social structure of the parishes investigated is relatively simple. 'The end to be achieved and towards which the society is organised, over and above the provision of food and shelter, is the preservation of property' (p. 242). He maintains that the society has a relatively simple economy and a rigid but culturally undifferentiated class structure which is based on property. But McNabb fails to highlight sufficiently how this 'simple' social structure only benefits the male members of the society. He presents a somewhat homogeneous impression of the society, saying that despite a rigid class structure based on property, it is culturally undifferentiated. But property allocation is based on a fundamental cultural division of gender roles, which further results in major social, economic and political differences based on gender.

## Inis Beag

John Messenger's anthropological study *Inis Beag: Isle of Ireland* (1969) is based on research he carried out on an island in the Irish Gaeltacht (Gaelic-speaking area), which he identifies by the pseudonym 'Inis Beag'. He and his wife spent most of a year there in 1959–60, and they returned eight times, with visits ranging in length from one to seven weeks, between 1961 and 1966 to complete their work. He described the main thrust of his study as 'the documentation of the contemporary culture of Inis Beag: its technological, economic, political, social, religious, aesthetic and recreational aspects' (p. 2). Standard ethnographic research techniques were used: guided and open-ended interviews, external and participant observation, collection of life histories, photography and phonography.

The importance of property ownership to relationships, social status and prestige on the island is clear from Messenger's analysis. Inheritance on the island is patrilineal and sons are given the farm by their fathers or widowed mothers. The prevalence of celibacy and late marriage is a matter of grave concern on the island, and the system of inheritance is intrinsically linked to these social patterns. The land cannot be sub-divided, and one son must wait until his father dies or is 'forced' to pass on the land, and his siblings have emigrated or married (p. 68). By the time this happens, sons are often middle-aged. A father who does not choose an heir among several sons will delay their marriage opportunities if they wait to see who will inherit. Messenger recounts how most fathers in Inis Beag are loath to surrender their property, and with it control of the family (p. 68). He describes how some fathers postpone passing on the land in order to maintain their status position past the age of seventy years (p. 68). The importance of property ownership to the father and sons is clear. It is a source of status and authority, and it is gender specific.

When the islanders were asked to rank prestige symbols and the positions of individuals in the status hierarchy, Messenger found that the only consensus which emerged was 'the placing of land and capital at the apex of the rank order of symbols, the assigning of men who possess these in greatest quantity to the top of the hierarchy, and the placing of landless men at the bottom' (p. 85). Messenger shows, too, that most of the major disputes on the island are over the ownership and inheritance of land (p. 82). There is much sibling rivalry on the island, one of the main causes of which is competition for the patrimony. Fathers sometimes play off sons for the patrimony and this leads to factionalism within the family. The status and prestige attached to property ownership is explicit, and so, too, is the desire of the men on the island to be landowners.

If a man is without sons 'he most commonly wills the land to a brother's son or arranges to have a man wed his daughter and "marry in" to the household' (p. 72). Interestingly, such a son-in-law may pay a bride wealth, which usually is much larger than the dowry of women, because he has no land of his own. Plainly this is linked to the man obtaining property through marriage rather than through inheritance. Throughout the study, Messenger always refers to landowners/ farmers as male and a very forceful impression of the island men's accepted control of and rights to the land emerges. The description of land ownership, inheritance, and land as a status symbol have clear implications for women. As non-owners, women are automatically excluded from the communally agreed status hierarchy since prestige is scaled according to the amount of land owned. This, and its implications for the role definition of the woman, are not investigated. Within the family, Messenger outlines how fathers seem to identify ownership of the farm with control of the family and are most reluctant to hand over both to their sons. Again this process totally excludes women, except in so far as she presumably occupies a position in the family over which her husband is reluctant to relinquish control. The way in which unusual situations are dealt with re-establishes the norm; widows, by virtue of having outlived their husbands, become landowners but they revert ownership back to sons on adulthood, and in the absence of sons, the son-in-law becomes the owner. For a moment women seem about to break into ownership patterns, but they quickly re-occupy their invisible positions as the land reverts through them to their male relatives.

Messenger's account further reveals the exclusion of women from social structures and practices, and also a measure of discontent with their situation. Women confided to his wife that they were unhappy about being forced to remain at home, minding children, and performing tedious household chores. They were resentful of their husband's greater freedom, and their involvement in numerous social activities 'forbidden by custom to women' (p. 77). Messenger also provides an analysis of power structures on the island. He distinguishes between formal control and informal control. He identifies the manner in which these structures define the parameters of social behaviour:

> formal systems are governments whose controls are based on physical violence or its threat – physical punishment, ostracism, banishment and extermination; while informal systems include sanctions of a religious or economic nature and those exercised by primary groups – gossip, ridicule, persuasion and opprobrium. Controls of a negative nature are buttressed by positive rewards, such as services provided by a government or the feeling of approval gained from primary group approval. (p. 55)

One interesting feature of the formal and informal social control positions he details, but on which he does not focus, is the absence of women. No woman occupies a position of formal or informal control. Women expressed concern about the pressure of informal controls, particularly with regard to having children (p. 77). Similar to Brody's analysis of Inishkillane (1973), Messenger reasons that many of the girls (*sic*) who emigrated from the island did so because they were dissatisfied with the lot of married women on the island (p. 125). Emigration as a way out suggests that the power of social structures determining roles on the island are such that enacting gendered changes was more difficult than emigration. While Messenger does not explore the root of the patriarchal nature of this society, it is clear that this lies in the ownership of property.

**Tory Island**

Robin Fox carried out his research on Tory Island in 1963. He clearly states in the introduction to his study, *The Tory Islanders: a people of the Celtic Fringe* (1978), that he does not attempt a complete ethnographic account of Tory Island, rather he examines several features of the social order: genealogy, kinship, inheritance of land, recruitment of boat crews, marriage and household (p. ix). His analysis leads him to conclude that there are many things about the Tory social structures which appear peculiar and are very different in form to those of mainland Ireland. Here, we focus on the system of inheritance.

There are some striking differences in the island life described by Fox, compared with mainland Ireland. He outlines the various status symbols in operation on Tory: good dancing, singing, being an outsider, receiving an 'official' salary (p. 28). Surprisingly, compared with the rest of Ireland, Fox does not identify ownership of land as being a source of status. However he does say that ownership of land is still a matter of importance for Tory people (p. 83) and 'is jealously regarded, its disposition often a matter of dispute and always of debate' (ibid.). Fox outlines the ways in which the Tory system of inheritance contrasts markedly with the general Irish system. On Tory every child of a landowner has a right to a portion of his or her land; 'there are no dowries, males and females can inherit land equally, and do' (p. 99). The principle which decides amongst the claimants seems to be the consideration of 'land of the marriage' (p. 106). That is, each marriage requires land and this must be provided for. If a woman or man were to marry, and their spouse had land, they would not press their claims to their home land but would leave it for the other siblings. The idea is that every household will end up with some land. Fox suggests that it is more correct to think of the islanders as 'holding' land rather than owning it, since relatives to some degree retain their claim on the land.

This system of inheritance seems, at first glance, to be quite remarkable. It is more egalitarian and inheritance is not sex-related. Fox describes it as a sensible 'human adaptation to a difficult environment' (p. 126), in that it allows for the rational distribution of land which, given the island's population and terrain, would not be facilitated by the patrilineal system of inheritance. Holding land is a more accurate description of the relationship to the land than ownership. There is no absolute right to property: rather, many members of a family retain a claim to the land. None the less, official land evaluations are more likely to credit the land as belonging to the man for valuation purposes, but deeper analysis shows this is not always correct. He then goes on to say that men will predominate in records of ownership since 'women are far more likely than men to relinquish claims' (p. 99). It is unclear why this should be so, if women and men are truly equal heirs. Fox says that 'a piece of land may be spoken of as "belonging to" a man, but will on analysis turn out to belong to him and two sisters' (p. 99).

Fox attributes the discrepancy between island accounts of land ownership and that of official records to the tendency of government evaluators to attribute land to men for valuation purposes (p. 99). He recounts how some cases of land recorded as belonging to a man, turn out to have 'come to him' from his wife, but have been treated as his by evaluators. Fox points out that evaluators are not worried about where land came from or where it will ultimately go. They are interested in recording the 'owner' who will pay the rates for the land, and this is usually the man who works and manages it, as far as they are concerned (p. 100). In effect what we see here is the state's difficulty in grappling with a system of land ownership different to individual private ownership. The way in which it deals with this is to impose a veneer of private ownership, although Fox argues it is a misrepresentation of the relationship between individuals and property on Tory. The state chooses to attribute ownership to men, in accordance with the legitimate ideology of patrilineal inheritance and male ownership of land. However, male dominance over land is not simply an artefact of mainland patriarchal ideology. Fox himself recounts that on Tory, men are more strongly identified with land in terms of ownership. He argues that men appear to monopolise ownership because it is usually they who work and manage the land. It does seem that while there is a different system of land acquisition, land is still largely identified with men, and Fox speaks of land 'coming' to a man through his wife. He does not speak of land 'coming' to a woman through her husband. While a woman may inherit land, it is ultimately understood as belonging to men. Men are more strongly identified with land in terms of ownership, and they are seen as the person with whom business should be transacted.

While the normal patrilineal system of inheritance of the rest of Ireland does not hold on Tory, holding/owning land is still seen as the legitimate domain of men. In this light, then, it is not so unusual that women are far more likely to relinquish their claims to the land than men: they do not consider that they are equally entitled to it. Fox describes the greater migratory pattern of women, particularly young women, off the island, which again suggests fewer ties to land and ownership of property. The normal pattern of patrilineal inheritance did not hold on Tory, and in terms of island survival, the system in place is eminently more sensible. Fox describes it as a rational distribution of land that is perhaps the best solution for this particular terrain and population. However, it has not significantly weakened the understanding of men as the legitimate owners/ holders of property.

**Inishkillane**

Brody's famous study, *Inishkillane: Change and Decline in the west of Ireland* (1973), is based on participant observation carried out in five communities in the west of Ireland between 1966 and 1971. He lived and worked in the communities as a visitor or additional hand, but never as an investigator. He also used information and statistics from parish records, Dublin offices of rates and land-commission, and information provided by key informants. It is through these channels that Brody came to believe that demoralisation was rampant in the west of Ireland; 'it is the breakdown of the communities, the devaluation of the traditional mores, the weakening hold of the older conceptions over the minds of young people in particular, to which every chapter will return' (p. 2). Brody declares that, unlike Arensberg and Kimball, he is not about to describe a harmonious and self-maintaining system, but rather one in which the people are demoralised and have lost belief in the social advantages and moral worth of their small society (p. 16).

   Brody describes various social changes which he presents as symptomatic of demoralisation: patterns of interaction in the bar, disillusionment with rural life compared to urban life, and a general decline in commitment to the traditional society. None the less, relationships to property, and patterns of ownership and inheritance remain intact. The interesting feature of Brody's study is that rather than presenting ownership of property only as a source of power and privilege, he also presents it as a responsibility which obstructs the life-choices an heir can make. Inheritance entails a responsibility to continue the family farm, to remain in the demoralised traditional society, and at one level Brody presents this as a gendered *disadvantage*.

Brody identifies the disappearance of mutual aid as a symptom of the change and decline in rural Ireland. This system, which was important economically and as a source of interaction, is being replaced by a strong emphasis on the privacy and self-reliance of each household. Brody argues that this results in a certain vulnerability for parents, since if a son does not stay to farm, they will be isolated (p. 123). He further argues that fathers are aware of the rapid disintegration of life in the west of Ireland, and in the face of this, it is more difficult for them simply to expect their own sons to stay on the land (p. 120). Throughout the transition to self-reliant households, though, the gendered nature of ownership and inheritance has remained intact. It is still hoped that a son will stay on the farm.

Brody presents emigration as a means of escape from a disintegrating society for disenchanted young people, and outlines the differing rates of emigration for men and women. Women leave when they are younger and they leave in larger numbers. In explaining this difference, he says:

> the woman's role and status in the social structure have, ironically enough, given her an emotional freedom to bring the most drastic change of all to the communities. Having no part in any inheritance, the woman has always been without any material possessions. Sons inherited the house, the very cups and saucers in it, and the land. But with the inheritance went duty and responsibility ... only with considerable difficulty could a son, the owner and inheritor, defy his duty and neglect his responsibility. Even the last daughter, however, has been spared this tension, and has felt free to leave home without guilt. (p. 127)

In this passage Brody indicates the way in which property provides different life-choices for sons and daughters. The son is seen as the legitimate heir, and 'son', 'owner' and 'inheritor' are used as synonyms. Because she is disinherited, the daughter is free to leave. Brody describes this outcome as 'ironic', presumably because her freedom arises from the norms that deny her access to property. However, Brody does not guide us through an analysis of how and why this is the case, how the situation mobilised against her initially and how it obviously remains intact if it is the cause of the extensive female emigration he documents.

The fact that it is easier for the last girl to leave than it is for the last boy (pp. 92, 93) reflects the persistent view of men as the legitimate heirs. Similarly Brody says that 'country girls have refused to marry into local farms' (p. 98), which shows that marriage is still the channel through which a female enters farming. He describes how men meet in the bar, which is the focus of community life, while the women meet in the shop. He says that those who are least involved in farming life are also the 'least likely to go to bars and least at ease with the bar's ways. By the same token, they feel least committed to life in the farming society' (p. 161). He says, too, that 'those who spend most time and feel most at ease in the shop are those furthest removed from farm work. It follows that they are also most removed from any commitment to staying in the countryside at all' (p. 163). However, there is a fundamental difference between being least committed to life in a farming community, and being most excluded by life in a farming community. The 'least involved' in farming, as described by Brody, are precisely those who do not inherit property. It would be more correct to say that those most at ease in the shop are those furthest removed from the power that comes from property rights, rather than those who are opposed to farm work. Farming structures, by tradition, do not acknowledge the contribution and value of female farm labour, but this does not mean that it is non-existent, or that they are removed from farm work. Rather, it is the work of the property owner that is formally recognised on farms.

Brody further details, without analysing, the exclusion of women from rural life. He says 'women were also excluded from the social centres of the community ... only when they reached old age was the wisdom attributed to years allowed to transcend, in public life, the insignificance attributed to womanhood. In home and community life alike, therefore, the woman's influence may have been significant, but it was informal and domestic: women had at least to appear to be without authority just as they were in practice without possessions' (pp. 110–11). This passage describes how women were excluded from all social centres, and were considered insignificant because of their gender, were propertyless and had to appear to be without authority. Yet all of these factors seem to be connected. In other words, the situation of women as described by Brody reflects the persistence of the gendered nature of property relationships. The high emigration rates of young women, which he highlights as a symptom of change and decline, are prompted by social structures that deny women access to property. Brody does not analyse the persistence of the patrilineal line of inheritance, and male ownership of property, which persists through what he describes as a disintegrating society. The only way around it for women is to leave.

## Culturally Prescribed to Negotiated Roles in Farm Families

Hannan and Katsiaouni's study, *Traditional Families?* (1977), holds a position of importance in the chronology of research on farm life. Their report makes the leap from the anthropological studies we have considered to the analysis of a modernised, commercial type of farming. Hannan and Katsiaouni state that their study is an attempt to provide some information on nuclear family interaction patterns in Ireland (p. 11). Their main aim is to identify the principle characteristics of farm-family interaction, explain variances in interaction, and examine how and why farm-family interaction patterns have changed in Ireland since the 1930s (p. 2). They believe the traditional family structure described by Arensberg and Kimball was typical of rural Irish society in the 1930s. They recount the main features of this as a clearly differentiated and deeply institutionalised division of labour, a definite patriarchal authority system with a simple routinised decision-making process concentrated in the father's role, and a distinct social–emotional role fulfilled by the mother whereby she acts as the main emotionally supportive and tension-management agent in the family (p. 21). While this farm-family structure existed and was suitable within a particular context, this context has changed dramatically and significant changes within the family structure are also to be expected. They identify two crucial processes which are accountable for this: the first is the commercialisation of farm production, and the second is the massive expansion of mass communication and modern transport. Hannan and Katsiaouni maintain that these forces combined are likely to lead to changes in people's beliefs and values 'as people begin to take on the perspective of prestigeful urban reference groups' (p. 26), and definite adaptations in family task and decision-making patterns will have to be made as a purely circumstantial response to the changing farm and household economy.

As modernising forces increasingly impinge on the traditional system, the expected direction of change from the traditional family structure is towards 'the modern urban middle-class' model, which is the other anchor point of Hannan and Katsiaouni's study. The summarised description of this model, developed by Elizabeth Bott (1971), recounts the main features as being minimal or no spousal segregation in housekeeping and child-rearing roles: similarly, power or authority gradients between spouses and between parents and children are minimised, with decision making being largely a joint consultative process. The greater openness of all interpersonal relationships within the family means that maternal specialisation in emotionally supportive functions is no longer obvious or necessary. The basic economic provider role is still predominantly male, and Hannan and Katsiaouni feel this will be particularly so on farms; 'in the farm situation, as we will make clear, increasing commercialisation almost necessarily exaggerates the male exclusiveness of the provider role' (p. 15). The traditional farm family and the modern urban middle-class model are the two anchor points for Hannan and Katsiaouni's study, and they set out to show and explain variations in farm-family interaction patterns along a continuum between these anchors. This study pioneers the recognition within Irish sociology of power relationships as a basic dimension of family interaction, and the necessity of attending to these power relations in order to understand patterns of family interaction. They also draw attention to emotional factors and spousal love as key components in family interaction.

There is little discussion of property or property relationships in Hannan and Katsiaouni's study. It could correctly be argued that it is not central to their work. None the less, it permeates the entire case they build. Hannan and Katsiaouni perceive the farm and household as completely separate spheres. The farm is clearly identified as the production sphere, the household as that of consumption. They observe that the increased separation of these has been instrumental in lessening or weakening the position of the wife in farm production. Her contribution to production is less essential now than it was previously. In the more economically rationalised households, she has become increasingly like her town counterpart, in that her role is almost exclusively confined to consumption roles, household and child-rearing tasks (p. 24). In their attempt to correlate the farm wife's role with the consumption role of the modern urban middle-class model, Hannan and Katsiaouni ignore her important contribution to farm production. They actually found that two-thirds of the farm men interviewed said that the contribution of wives and younger children to farm production was extremely important, and that without it some lines of production would have to be dropped. Similarly, Hannan and Katsiaouni say that a high level of economic activity requires a very high level of efficiency in farm, household and farmyard operations and in financial management generally (p. 160). This also reflects the importance of household activities for general farm production. In other words, while an urban middle-class model is superimposed on farm families, it obscures the differences of the farm household from the urban household, and the way in which women on farms are tied into production, while remaining disassociated from the means of production.

Immediately observable similarities with the modern urban middle-class model are emphasised, while the very different power dynamics these ostensible similarities may represent are played down. Hannan and Katsiaouni equate the 'necessarily' exclusively economic provider role of the commercialised farming husband with that of his urban counterpart. This continues to mask the productive importance of the farmwife for the farm business and it also shrouds other dissimilar power relations which occur in the farming context. Hannan and Katsiaouni point out that unlike most non-farmers, the work context of the farmer is very much within the family (p. 86). What they neglect to consider, however, is the probable effect that farm business and status positions occupied by individual family members will have on inter-personal relationships.

The persistence of the patrilineal line is given little attention by Hannan and Katsiaouni. In 79 per cent of the couples they interviewed, the husband had inherited the farm, while in only 8 per cent of cases had the wife inherited, but they do not reflect on this point at all. Indeed they gloss over it, saying that the 'process of land acquisition through inheritance and purchase is related to age and family cycle stage' (p. 69). In this way they neglect to draw attention to and examine the more pertinent relationship between land acquisition and gender. In the vast majority of cases, the 'farmer' owns the farm his father owned and which, in time, his son will own. This predominant patrilineal system of inheritance in rural areas again suggests latent gender power relations. The patrilineal line ensures men's position as owner, and women continue to be disinherited. Again the implications of this for spousal and familial interaction are not explored, nor is the fact that it continues within what they argue is a society fundamentally different from the one described by earlier anthropological studies. The functionalist framework employed by Hannan and Katsiaouni also seems to have hindered an analysis of how property relationships are intrinsically linked to the type of division of labour that develops.

Hannan and Katsiaouni do not consider the actual continuation of gendered property relationships with their modern urban middle-class model. Indeed, if this model develops in rural areas it seems to consolidate existing relationships, and the unequal power relations between the occupants of the alternative roles. Hannan and Katsiaouni assume it is natural for their 'modern urban middle-class model' to develop in rural communities. They say that the direction of change is 'almost inevitably'(p. 16) towards such a model. If there is an inevitability to such a development, it is intrinsically tied to the male control of property. Hannan and Katsiaouni identify how the survival of a given system relies on the belief that it is legitimate. They say of the traditional farm structure that such an overall system could only remain intact so long as it remained legitimised by the consensual sets of beliefs and values of the community. This legitimising ideology remained effectively isolated from contending ideals of family organisation which hold in external prestigious groups (p. 20). They argue that this traditional society no longer exists, and has now moved towards the modern urban middle-class model. Yet this model, with its emphasis on an exclusive male economic provider, actually legitimates and reinforces the structure of power and gendered property relationships that were present in the traditional system. They say that 'although the degree of participation by the husband in household and child-rearing tasks is limited by his economic role as provider, what is important is that the norms have changed' (p. 27). But the legitimate spheres of work have stayed the same. There has been a reinterpretation of the old pattern which allows it to remain acceptable. By stressing the 'male economic provider role', the farmyard continues to be legitimated as the male domain. Similarly, the patrilineal system of inheritance is justified. Hannan and Katsiaouni also neglect the manner in which increased contact with external reference groups may actually reinforce the fundamentals of the traditional structure. They argue that the participation of husbands and wives in formal organisation membership, in the mass media, and the 'increasing collaboration in a market economy which would involve the husband–father in ever wider networks of market relationships and his wife in more consumption orientated relationships' (p. 91), will expose them to meaningful alternative ways of organising family roles. But the depth of the alternatives provided by this external reference group is questionable; farming groups reaffirm the industry as a male preserve. It continues to be men who are involved in wider market relationships while it is women that are involved in consumption-orientated relationships. Thus contact with external reference groups need not signify fundamental change. They certainly provide little reason to question the fundamental gendered nature of property relationships.

Basically, Hannan and Katsiouni argue that change is afoot which is leading to more egalitarian power positions within the family. None the less, this continues to be underpinned by exclusive male access to property, and consequently to control of the means of production. Within farming then, 'farmers' politically continue to be represented for the most part as male, and socially and culturally the farm work of women is not immediately, or easily, recognised. The 'changes' which Hannan and Katsiaouni identify, should be more accurately described as 'adaptations', which leave intact the legitimacy of male access to land ownership.

## THE MIDLANDS

My own research was conducted in the Midlands during 1987 and 1989 (Shortall, 1990, 1992). The main focus of the study was to identify the various factors which render women invisible in farming, and why women continue to subscribe to such a situation. The main argument advanced was that an analysis of power is key to understanding the lack of a greater expression of grievance by women about a context that so clearly disadvantages them. It was a qualitative study, involving participant observation and interviews with twenty women, married or widowed and living on farms, as well as interviews with members of farm organisations, farm advisory services and local organisations. As for all the other studies reviewed in this chapter, I only focus here on inheritance and property ownership.

Two-thirds of husbands had inherited farms, and the most common reason for a husband not inheriting was having an elder brother who was the heir. There were always unusual circumstances in the cases of the very small number of women who inherited land; either they had no brothers, or in one case a woman had looked after elderly relatives for twenty years who eventually left her the farm. Evidently it is still not the norm for women to inherit land. Interestingly, where women had inherited land, their husbands became installed as 'the farmer' after their marriage, and all mail, information, cheques, and callers to the farm, came to him. This reinforces Salamon and Keim's (1979) argument that even where women do have access to land, it does not mean a straight reversal of power positions.

Almost all of the women interviewed had children, and each of them said that a son would inherit the farm. The strength of the patrilineal pattern of inheritance is obvious when we consider the age of the children. Four of the future heirs were under twelve years of age, three were under five, and in one case the couple had a son of two years and a daughter three weeks old, and still the decision had been made as to who would inherit the farm. This verifies the fact that heirs, those who will ultimately own land, are primarily chosen on the basis of sex. Some of the women spoke of daughters who were very interested in farming, and loved being involved with the farm. However, there was no consideration of their daughters as heirs, and the only future for their daughters in farming that was discussed was marriage to a farmer.

The patrilineal line of inheritance is a powerful social custom which advances the belief that men are the natural heirs to land. It is generally accepted by men and women as natural, and is so prevalent it is little questioned. The whole public world of farming deals with the landowner, and in this way reinforces the idea that it is an individual who farms, thus rendering women and their work invisible. Mail, information, cheques and callers to the farm all address the man on the farm. The way in which these props, again so taken for granted that they are almost beyond question, support the idea of men as the key figure on farms, and the way in which this is intertwined with land ownership, is evident when we consider the case of widows. Widowhood is something of an 'abnormal time' in the sense developed by Steven Lukes (1974), in that usual power structures are relaxed or diminished. Let us consider the case of the two widows in the study.

Upon widowhood, the women became the sole owner/controller of the farm. They assumed direct control of the farm, and were primarily responsible for farm decisions. This had not been the case before widowhood. One woman said, 'I make the decisions, although now that [son] is older, I discuss things with him more.' The other woman reported, 'I am very involved in farm decisions, although I was even more so at the beginning of my widowhood.' Both women have sons in their early twenties. Both reported that callers to the farm asked to see them, and mail, information and cheques were now addressed to them; 'The cheques used always be addressed to [husband], now they either come to me or my son.' One woman attends farm meetings which she had never done before her husband died; 'I am the only woman who is ever at the meetings, and I am not afraid to make a contribution either.' This last comment shows the woman's recognition that her situation is unusual.

The important point is that many subtle processes reinforce and recognise widows' changed power and status position when they take over ownership. In the 'normal' situation, all of these processes operate to reinforce the position of the male farmer. Interestingly, while farming structures recognised and supported the widows' changed power and status position, this was being withdrawn as their sons became older.

The patrilineal line of inheritance continues to deny women access to land ownership. Newby et al. have argued that ideologies of property ownership contribute to a system of 'natural' inequality in the country-side which can remain an extraordinarily prevalent feature of the taken-for-granted perception of rural society. This is certainly true of the patrilineal line of inheritance in Ireland throughout this century.


CONCLUSION

To reiterate Kennedy, it is remarkable that the kinship-impregnated institution of inheritance has persisted in such robust form into the late twentieth century. Such continuation is exactly what this chapter illustrates. Tracing through the main studies of farming from the 1930s up to the late 1980s, the strength of the patrilineal line of inher-itance, and the social norms and customs sanctifying this system of land transfer, are evident. The acquisition of land is shaped by social norms and traditions that clearly disadvantage women.

It is evident throughout the chapter that the acquisition of property is not simply about acquiring an economic resource. It shapes social relationships, and relationships in the wider community. In relation to the family, the reluctance of men to part with ownership and control of property shaped family structure for a good part of this century. The production of a male heir often determined the success of a marriage (Gibbon and Curtin, 1978; 1983; Varley, 1983; Curtin and Varley, 1984). With regard to the wider community, being a landowner provides access to certain fora, and owning land in itself is a source of prestige. The studies reviewed illustrate the different social position of women and men on farms, which is related to land ownership. Interestingly, most studies attend to the relationship between father and son with respect to land ownership. The gendered nature of land transfer and ownership is less of a concern, even though it is the single most important factor shaping the lives of women on farms.

The essentially individualist notion of property ownership which prevails means that the concept of the family farm is difficult to grapple with, and it invariably refers to a family living on a farm owned by a man. Fox's study of Tory Island illustrates the difficulties for official records of non-individualistic systems of land ownership. In the case of Tory, officials circumvented this difficulty by naming a man as the owner, even though many other members of the family had a claim to the land. The policy they followed, however, was in line with the traditional accepted norm on mainland Ireland.

There is no doubt that the status and prestige attached to land holding has changed throughout the twentieth century. Ireland has industrialised and modernised and the value of education has increased. None the less, these processes have not changed the mode of entry to farming. It continues to be through the inter-generational transfer of land, and this chapter illustrates that this continues to be a transfer between men. The wider societal changes that have occurred in Ireland have not affected this process. The status and prestige attached to farming have changed, and owning land no longer affords men the same status within the family and community. However, we have noted that landholders continue to operate as a status group, and this self-perception has influenced the greater social mobility of the children of small landholders compared to the working class. Property owners, even with a low income, continue to have higher aspirations and perceive fewer barriers to social mobility (Hannan and Commins, 1992). In addition, the symbolic importance of being a landholder and its status implications still count, and Hannan and Commins suggest it counts for more than the purely economic returns of the land (p. 84). In summary, the tenacity of male ownership of land has persisted throughout the twentieth century, and this continues to fundamentally shape women's role in farming. Individualistic property ownership is tied to an individualistic under-standing of farming as an occupation. The persistence of male owner-ship of land means that men continue to dominate with regard to the economic power, status and political power that accompanies land ownership and farming.

# 5 Farm Women, the Commercialisation of Dairying and Social Change

## INTRODUCTION

In this chapter, I will consider the commercialisation of the dairy industry that happened between the mid-nineteenth century and the early twentieth century in western Europe and North America. Farming throughout Europe and across the western world is marked by diversity. This is partly due to such factors as climate, soil, and distance from centres of consumption, but also to the fact that agriculture is a social construction. As Long and van der Ploeg (1994) note, the way agricultural practice is organised depends heavily on the actors involved in it, and the different social, economic, cultural and historical relations in which it is embedded. Long and van der Ploeg gently remind rural scholars that the behaviour of relevant actors interacts with the actions of the state and institutions to create particular and diverse outcomes. What is remarkable about women's role in the nineteenth-century dairy industry is the uniformity of change that occurred in different countries. I argue that this was primarily the result of their relationship to property, and their limited resources to successfully organise around their grievances.

International historical research illustrates that until the last century, women were traditionally responsible for dairying. Between the mid-1800s and the early twentieth century, women ceased to be the primary dairy workers, and men took over responsibility for the industry. For example, the research to which I refer in this chapter on Canada, the USA, England, Ireland,[1] and Denmark all testify that until this period women were responsible for the dairy industry. Equally, over about seventy years, this pattern had utterly changed in each country.

What happened? Why did women cease to be the key people in the dairy industry, and why did men assume responsibility for this area of work? As this chapter unfolds, and pulls together research on this question from the countries mentioned above, a number of patterns develop. Firstly, the importance of the wider economic context is clear. The significance of dairying changed, and this subsequently affected the gendered division of labour. Secondly, the prevalent gender ideology of the time influenced the roles and aspirations of women on farms. Victorian ideology seeped into farmyards and farm households, as well as influencing the role of women in the public space of farming. Thirdly, the role of the state in moving women out of dairying, and moving men in, is obvious. This is tied up with the first and second points; given that the state is capitalist and committed to capital maximisation, it became a priority for the state to develop the dairy industry as its economic significance grew. It was men, not women, who controlled land and capital and the public space of farming, and thus it was imperative to have men committed to the development of the dairy industry. Fourthly, and this is where Long and van der Ploeg's reminder is important, it is clear that the changeover is not just a simple story of state and institutional direction. The picture is coloured by the action and reaction of farm women and men. Of course women did not all respond in the same way. Some were pleased to be relieved of the drudgery of dairy work, others resented and protested at the loss of a skilled and prestigious occupation. The resistance looked at in this chapter was visible, but it is also likely that a great deal more resistance took place that is less visible, at least to the historian's eye.

It is important to be clear about why, of all the varied work that women did during this period, the focus is on their work in the dairy industry. Women have worked hard in farming for a long time. It was expected that women carry out essential productive work. However, as Faragher (1981) noted, the question of status for farm women is not primarily a question of what they do, but rather the recognition they are granted for what they do. He argues that despite the essential work farm women did, there is little evidence to suggest that their husbands and sons granted them equal power for equal work. This is undoubtedly true. Yet dairying was valued work. It was one area of work where women did receive recognition, status, income and a certain degree of power. It was unusual in many respects, and to move or be moved out of this field had particular significance for women.

The economic power of property ownership is very clear throughout this chapter. Although women had responsibility for dairying, they did not control the resources of the dairy industry. For a capitalist state committed to capital accumulation, it was logical to ally with those who had the economic power to invest in the industry. The actions of the state were easily legitimised by reference to the Victorian ideology of the time. While some examples of resistance are outlined throughout the chapter, women in dairying were organisationally outflanked; they did not have the resources or organisational power to resist with any significant consequence.

The chapter begins by detailing the historical accounts of women's role in dairying in a number of countries. The recognition women received for this arduous and frequently tedious work is sketched. The movement of dairying into the male domain of work was orchestrated. The active role of the state is discussed. The state's actions, to employ the distinction made by Alexander (1987), were both rational and non-rational. Rationally, in order to maximise efficiency, it was essential to engage with those actors who could invest in the industry. Non-rationally, the state's actions were influenced and constrained by the normative and moral beliefs of the time. Managing dairies, and possibly men, and generally occupying prestigious public roles, was not the norm for women. While the state did not create this ideology, it certainly reinforced its existence in dairying, and called on it to legitimise its own course of action. As the transition in dairying is examined, the reaction of farm women comes through. Many were happy to move towards a more home-centred position; others vehemently opposed the undermining of their role. This chapter focuses more on those women who resisted. This resistance illustrates some of the aspects of power discussed in Chapter 2. They were organisationally outflanked, not having the structures or means to give strength to their protest. Their position within dairying was insecure as long as they did not control the resources of the industry. None the less, these women's actions constituted expressions of grievance and protest at an exercise of power which shaped their lives. A key question, which ties in with a later chapter, is the form of resistance undertaken. Women did organise in groups, but usually as auxiliaries of men's farming organisations. In other words, their organisational powers were limited. Is it true, as sometimes argued, that collective activity is more difficult for farm women because they do not appear as a collective workforce? These are the questions that will now be teased out.

## WOMEN IN CHARGE OF THE DAIRY

Research on all of the countries we are about to review note the difficulties of dating the changed nature and gender of dairying. This largely reflects the reality of the transition; it occurred at different times in different countries, and in different regions within countries. In general, the period of interest is between the early/mid-1800s and early 1900s. What was the position of women in dairying prior to these changes? Let us begin with England.

## England

Ivy Pinchbeck's classic work, *Women Workers and the Industrial Revolution 1750–1850*, was first published in 1930. At the time such a detailed study of women's work was unusual. It is an invaluable source, and provides a vivid account of the harsh work undertaken by women on farms in England. Pinchbeck details the different work carried out by women from field work, harvesting and husbandry to dairying, describing the different lot of women of different social classes. Over the period of her study, the farm work of women had completely changed.

Pinchbeck identifies dairying as the most important and productive branch of women's work in agriculture in the eighteenth century. It was such responsible work that, on all but the largest farms, the mistress supervised dairy maids through every stage of the business and performed all the more difficult operations herself (p. 10). The dairy woman was not only concerned with the actual making of butter and cheese. The farmer depended on the advice of the dairy woman on the sale and purchase of stock. In addition, the rearing, feeding and sale of calves were entrusted to her (p. 11). Scientific principles were unknown in the eighteenth century, and successful dairying depended on the accurate judgement and experience of dairy women, gained from daily practice and minute attention. Hence the mistress of the house was herself responsible for cheese-making, and on a bachelor's farm, the dairy woman ranked as a mistress and lived under the most favourable conditions. As well as the dairy mistress, the dairy maid was also considered far more respectable than the common field labourer. Her work was considered to improve her as a prospective wife and mother (Kitteringham, 1973).

The work of the dairy woman was the most skilled of all the work carried out by women. It was also the most financially remunerative, with the prosperity of a dairy farm depending almost entirely on women's work. Pinchbeck states that the farmer's wife who managed a dairy made a considerable contribution to the farm income.

While the work of the dairy women was the most skilled, it was also without question the most arduous (Kitteringham, 1975; Pinchbeck, 1981; Davidoff, 1986). Women's work in cheese dairies often began at three or four in the morning, and they often worked until as late as ten in the evening, with the hours in a butter dairy being somewhat shorter. Pinchbeck describes in greatest detail the long days of dairy women, and the frequently boring and hard nature of their work.

In the late eighteenth century, there was a shift from dairying to arable agriculture, and dairying became less important while grazing became more so. This change was a response to the market; the prosperity of London meant that there was now a greater demand for veal, and farmers found it paid to give up dairying and take to grazing. In the process, women's dairying, which had taken place on almost every farm, was slowly phased out. In the purely dairying districts of the north and west, a new process developed whereby dairies were let out. A new group known as dairy men appeared and rented the dairy. Women's trade was transferred to these men, who devoted themselves entirely to the task in hand. The wives of the dairy men had a more limited sphere than previously. They were not the controllers of the business, but were confined to the making of cheese and butter alone.

Pinchbeck presents the transition of dairying to a male occupation as one that was wholly favourable to the farm wife. While she sacrificed her former economic independence in so far as she ceased to contribute to the wealth of the family, the new arrangement meant an advance in the social scale and social activities of women, without entailing any material hardship (1981, p. 42). The fact that women no longer had to undergo strenuous practical training for their dairy work meant that they could now undertake an education fit for social activities. Davidoff also points out that the wish of women to move beyond their farm work should not be dismissed as sheer snobbery. Their physical labour was grinding, and of a nature that left farmers' wives socially isolated.

Research on this period gives us a sense of the moral norms and social customs which provided the context for this transition. In the 1840s, agricultural work was condemned as too arduous for women (Pinchbeck, 1981, p. 102). Interestingly, this concern was inconsistent; there was a great deal of concern about the unsuitability of women's field work and work in gangs, but little about dairy work which was undoubtedly more intense and physically demanding (Pinchbeck, 1981; Davidoff, 1986). It is unclear the extent to which the dominant ideology motivated social change, or whether under-lying economic reasons which prompted the gender shift in agricul-tural work were explained and legitimised by reaching for prevalent normative beliefs.

Farm women had to balance the prevalent Victorian ideology concerning the spiritual and material rewards of domesticity against their direct contribution to productive farm work. As Davidoff (1974; 1986) explains, other social norms restricted women's role in farming. The public house was the usual place to transact all types of business. Respectable women of the time did not go to public houses, as evan-gelical opinion associated them with sin. In addition, women rarely controlled large sums of capital, and were not seen as legitimate borrowers for investment. After dairying became men's work, the visible, acknowledged economic role of women in the farm was dimin-ished. This transition tallied with the ideology of the time, which frowned upon women's appearance in the public space of farming.

## Ireland

Men assumed responsibility for dairying at a later time in Ireland than in England. Joanna Bourke has written extensively on this period in Ireland (Bourke, 1987; 1993). Again it is difficult to date precisely, but it was nearer to the turn of the century before men assumed responsibility for the industry. In Ireland the transition is closely asso-ciated with the development of creameries. As in England, dairying was both valued and arduous work, and it was predominantly women who were in charge of the industry. At the end of the 1800s, the price of butter rose and remained high. In addition, mechanical churning and separating was starting to supplant home butter-making, and the conversion of a domestic handicraft into a more or less highly cap-italised factory industry had begun (Horace Plunkett Foundation, 1931). The dairy industry was vital to the Irish economy, and by the early 1900s, the industry and female participation in it were both changing. The dairy industry grew rapidly, and in Bourke's words, women were 'driven out of the industry' (1993, p. 81). It is this process that will now be considered.

Milking was the first dairying job to be taken over by men. Where milking was for the household, women were more likely to milk, whereas if milk was supplied to the creameries, it was more likely to be undertaken by men, or men servants (Bourke, 1993, p. 84). The cream-eries were introduced to regularise production and marketing, and to ensure uniform, high-quality butter. While creameries were economically desirable because of their efficiency, they lowered female employment and significantly reduced the income potential of women and girls (ibid., p. 87). Given that women had traditional responsibility for the industry, this employment pattern represented an important change. One of the main reasons given at the time for the preference for male employment in creameries was technology. Women were not considered strong enough to handle the heavy machinery. However, as Bourke notes (p. 90), even in those managerial posts which required no physical strength there were few women.

Interestingly, men were not initially interested in assuming responsibility for dairying. They perceived it as women's work. In the 1880s a man who knew anything about dairying was viewed almost with contempt (ibid., p. 99). The state took an active role in encouraging men to play a key role in dairying. Agricultural class books had encouraged boys to read the chapter about dairying since the 1840s. A number of state and private bodies provided dairying and training courses for men from the 1880s; a course for creamery managers was started by the Department of Agriculture and Technical Instruction (DATI) in 1895, largely attended by men. In 1899 it was decided to discontinue all courses for women, on the grounds that it was too expensive to run a course in dairying for women and one for men in separate halves of the year, and that it also led to inadequate training for men (ibid., p. 96). There was also a great deal of ambivalence about continuing to train women in dairying, as it was argued that many students were using their training as a stepping-stone to emigration. The training of men in dairy management was considered advisable given that 'intellectual direction' of the dairy was necessary. Even though the dairymaids in the creameries had a greater knowledge of the industry, the Department followed a policy of pushing men into managerial positions (ibid., p. 99).

In the early 1900s, female unemployment caused by the advent of the creameries was of some concern. The ideology of domesticity was also invoked in Ireland in this context. It was argued that churning had been inhuman work for women, and caused premature ageing. Housework and poultry were offered as alternative and more suitable work for women. Increasingly, women were trained to be rural wives, largely through schools of rural domestic economy. Another significant restriction on women were the hours of work enforced by the Factory Acts at the turn of the century. Amendments which did allow women to work on Sundays only allowed them to work for three hours. As a result, many creameries replaced their female dairy maids with male operators, as it was impossible to get through the necessary Sunday work in three hours.

Those women who were employed in creameries resented being managed by less competent men. The relationship between managers and dairymaids was frequently antagonistic (ibid., p. 99). This generated some public debate at the time in magazines, between defenders of the male managers, and sceptics who argued that the reason for the antagonism was the greater skill of the dairy maids, and the familial and political appointment of male creamery managers. The resistance of women reflects a general awareness of their level of skill. While the friction between the dairy maids and the creamery managers may have amounted to little more than a thorn in the managers' side, it does reflect discontent at an exercise of power, and a decision imposed on women of the time. The resistance of women in Ireland to the proposed reform of the poultry industry during the same period was more successful than their protests about changes in the dairy industry, and was partly responsible for the delay in significant changes to poultry-rearing practices. The state also tried to engage men in the industry, but because of women's resistance, the limited value of the industry, and the fact that it did not become a capitalist industry until much later, women retained control (Bourke, 1987).

**Denmark**

There are noticeable similarities between Hansen's (1982) account of dairying becoming a male industry in Denmark and Bourke's account of Ireland. Prior to about 1880, dairying was women's work, predominantly carried out by dairy maids and farm wives. From the 1830s, there was great affluence in Denmark, based on the market for grain which commanded a high price. Around this time, there were also some fears about the dangers of monoculture, and it was decided to combine grain with cattle-raising. This increased the emphasis on dairying and the importance of the activity for women.

In 1836, the Royal Danish Society of Land Economy was requested to undertake the training of women to be in charge of butter production on a larger scale than ever before (Hansen, p. 227). This was a new departure in education for women, and in some cases, women undertook two years' education. As Hansen notes: 'One must conclude, therefore, that the Royal Danish Society of Land Economy was confident that women could be in charge of independent work of great economic significance for agriculture' (p. 228). Although there were few large estates, those dairy maids employed on them obtained extremely independent positions and very high wages. In some instances, skilled dairy maids made more than men. The women had huge capital values under their control. However these women were few, and it was usually the wife of the house who had the key position in dairying. Hansen argues that while women in Denmark had few legal rights over the farm, their role in dairying detracted significantly from this restriction. Women were confident of their skills and the value of their work. Some farmers' wives organised training of younger women for a small fee, and dairy maids published directions on butter- and cheese-making. It was unusual for women to write publicly at this time, but clearly women were aware they had a skill in demand and were qualified to elaborate on it (p. 229). Hansen maintains that during this period women had achieved a highly recognised position because of the economic rewards of their work in dairying, which led to a hitherto unknown degree of self-confidence and authority (p. 223).

From the mid-1870s, the economic situation of farming changed. The price of grain began to fall, and it was realised that grain had to be abandoned as an export commodity and replaced by dairy products. Co-operatives were formed to overcome the problems of expensive equipment, and to surmount problems of the uneven standards of produce. Because women had charge of dairying prior to the advent of creameries, it might be expected that they would continue to be in charge of production in co-operative dairies. However, this was not to be the case. Women were, in Hansen's words, 'forced out of their leading role' (p. 234). Most dairies hired a man with some knowledge of machinery as a dairy manager, along with subordinate female personnel. Following on from the changed status of dairying, agricultural schools offered training for men as dairy managers, and as soon as trained men were available, they were hired. Hansen notes that the rationale for the exclusion of women was not based on an assessment of their skill, but rather on appeals to appropriate gender roles; it was put forward as unnatural to see a woman as a manager of a large operation, as they would be unable to negotiate with male operators, and they lacked physical strength (ibid.). Hansen also notes that there was a fear that women would work for lower wages, and would thus depress wages generally. Again we are left wondering to what extent the prevalent gender ideology dictated certain patterns of behaviour, and to what extent it was invoked to advance a situation desired by the state or male workers.

Quite a lot of discussion took place between men and women in agricultural journals regarding the qualifications of the sexes for dairy production. Many women resisted the new turn in their dairy work. They argued that men should allow women to decide for themselves what their work should be. They also rejected the idea that their lack of physical strength would hamper them in their work, arguing that machines had made the process less physically demanding. They argued that as women had brought the occupation to its leading position, there was no reason for them to leave it now (Hansen, p. 235). Despite their protests, women worked in subordinate roles in dairies, and their low-paid work lost all of the prestige previously connected with the dairy maid. One other form of protest is worthy of report: women boycotted menial tasks on the farm. Although the excellence of weeding as an occupation for women was promoted in the 1880s and 1890s, indigenous women did not undertake this work and immigrant women were employed instead. The increased self-consciousness of women meant they would not accept unskilled, menial work providing little remuneration (Hansen, p. 237). In Denmark, as in Ireland, the developing trend was to promote domestic economy education for women.

## Canada

Marjorie Griffith Cohen (1984) provides a very sophisticated analysis of the decline of women in Canadian dairying in the late nineteenth century and early twentieth century. The development of dairying in Canada had more in common with the United States than with Europe. There was little dairy specialisation, instead it was one of the many activities carried out on the farm. Less concentrated land ownership, limited markets and a scarcity of wage labour in the New World gave dairying a different appearance (Cohen, 1984; Faragher, 1981). In Canada, the work of the dairy maid was less likely to be a distinct, skilled occupation. Rather, dairying was only part of the labour of a general domestic servant, and more usually, it was performed by the farm wife. None the less, dairying was a valued activity on the farm. Cohen reports that accounts of the time indicated the importance of dairying to the whole farm operation, and recognised that when dairying generated income, it was the result of female labour (p. 309). It is also clear from comments of women at the time that they regarded the income generated from dairying as belonging to them, even though the proceeds were often applied to the family's keep.

Dairy equipment was primitive until the late 1800s. This was not because of any scepticism on the part of farm women, but rather because they did not control capital expenditure on the farm. Men did, and as dairying was rarely the primary focus of farm operations, there was a tendency to ignore it when capital improvements were considered (Cohen, p. 313). However, the economic conditions of farming changed dramatically, and from the mid-1800s, the market for dairy products grew considerably. This was the result of the opening of US markets, where the price of Canadian dairy produce was very competitive in the wake of the American Civil War, which caused considerable disruption to domestic production. In the 1860s, the export of cheese and butter to the United States increased dramatically. With the prospect of high prices for dairying, combined with uncertainties in wheat production, many farmers were tempted to specialise in dairy farming. Very rapidly, cheese-making moved to the factory, and large dairy herds were developed to supply milk to the dairies. In the process, dairying ceased to be a part-time occupation of women, and became the major work of men on farms (Cohen, p. 320). Women retained control of butter-making much longer, as it continued to be made on the farm. The transition to the factory, and a changed role for women, did not occur until the turn of the century.

The state played a key role in promoting dairying as men's work. It stressed an important change in the nature of dairying: it became a 'scientific' occupation and therefore more worthy of the attention of serious farmers than it had been before (Cohen, p. 331). Women's dairying, on the other hand, was presented as an instinctive sort of process. The serious business of dairying was presented as one that needed the commitment of men. Educational promotion was an important aspect of government efforts to stimulate dairying among male farmers. As Cohen rightly points out, realistically there were few ways the government's actions could have been different. The promotion of dairying as a male activity reflected the economic position of women, and the fact that if there were going to be investment in the industry, then it was necessary to engage men on farms. Men owned the land and controlled capital expenditure. Furthermore, the norms and beliefs of the time were such that a more capital-intensive industry outside the home was seen as rightly the sphere of males. Still, government aid accelerated the decreasing participation of women in dairying.

Cohen notes that while some women undoubtedly welcomed the reduced work-load that resulted when dairying ceased to be their responsibility, for others it represented a certain loss of independence. While many women continued and still continue their involvement in dairying in Canada, the perception of their role has changed completely, to one of mere assistance.

## The USA

In the USA, as in Canada, dairying was one of the many tasks for which women had responsibility. It was not specialised work; women made butter and cheese, tended vegetable gardens and were generally self-sufficient (Jensen, 1981; Sachs, 1983; Osterud, 1993). Nineteenth-century accounts illustrate the remunerative importance of women's butter-making in southern, eastern and mid-west farms. In a similar process to Canada, men displaced women as dairying became commercialised, and marketing techniques were centralised (Jensen, 1981). One possible explanation for women's lack of resistance to their removal from dairying is that they welcomed such a move. Like Pinchbeck, Jensen details the drudgery of dairy work, and its long, arduous nature. Osterud speculates as to whether the displacement of women by men reflected a male assertion of authority, or women's withdrawal from a laborious task. It is more likely that it reflected the fact that dairying was women's occupation as long as it required little capital investment. Large-scale capitalist enterprises did not grow from women's work (Jensen, 1981). It became men's work at least partially because they controlled the investment that would make it large-scale and capitalist. Bokemeier and Garkovich (1987) maintain that American women were also constrained by the cult of domesticity prevalent in the USA at the time. They refer to the fact that women's farm work was often redefined as housework for precisely this reason; if farmyard tasks were perceived as *housework*, it was permissible for women to undertake this work. Thus in the United States, too, socio-cultural values and goals provided the context within which gender roles were devised.

What was unusual about farm women in the USA was that they were solicited to become involved in agrarian protest organisations. The National Grange of the Patrons of Husbandry, more commonly known as The National Grange or the Grange, was founded in 1867 to alleviate the financial difficulties of their members by establishing co-operative stores, purchasing agencies and factories for the manufacture of farm machinery. It was also a social organisation, arranging social and community events. The Grange admitted women to full membership, reserving four offices for them while excluding them from none of the others (Marti, 1983). The National Grange refused to charter any state organisations that did not follow these rules.

The Farmers' Alliances peaked in the 1880s and 1890s. Their protests reflected the frustration at the decline in the standard of farm living, and the steadily falling prices for farm products. The Alliances were eventually displaced by the Populist Party. The Alliances, like the Grange, also allowed women full membership but unlike the Grange, did not reserve offices for women, although women were occasionally elected to office (McMath, 1975).

These instances of women playing a leadership role at a local and national level in farming organisations was quite remarkable for the time. The Alliances and the Grange provided women with an opportunity to share political activity with men, and it is argued that the organisations were pioneers of a mixed sex political culture (Jensen, 1981; Watkins, 1993). They provided a forum for women to practise and advocate women's rights. Both the Alliances and the Grange debated suffrage, the Grange endorsing and withdrawing endorsement twice, before finally approving it in 1915. Women, too, were divided on the question of suffrage. This was reflected in the ambivalence of both organisations to women's equality, while formally espousing a commitment to it. The public leadership of the organisations remained the preserve of male members. Some radical and exceptional women did exercise leadership in the organisations, but they did not translate their vision into terms farm women found relevant (Jensen, 1981; Marti, 1982; Watkins, 1993). It is argued that one of the reasons the organisations advocated women's membership was because they needed the strength of numbers (Jensen, 1981). While the Grange demonstrated its ambivalence about suffrage, a strong argument in favour was the organisation's commitment to temperance. This commitment, combined with the perceived conservative views of women on temperance, was one reason for advocating suffrage.

Women were usually involved in the social activities of the organisations, occasionally doing some lecturing, and working with children. Women's comments on agricultural matters were mostly limited to poultry and dairying, which were seen as within women's sphere. In the late 1880s, Granges organised to protect the butter market. Through their Granges, women and men put pressure on Congress, forcing it to pass stiff regulatory laws controlling the manufacture and sale of oleomargarine, a butter substitute. The law introduced succeeded in reducing the sale of oleomargarine for about a decade.

While this collective action relates to women's dairy work, it happened prior to any change in the gendered nature of labour division. Indeed, it is unimaginable how questions of the gender division of labour on farms could be raised through farm-family organisations.

## DISCUSSION

Despite the different cultures and systems of agriculture considered in this chapter, a remarkably similar pattern emerges in the transformation of dairying into a male industry. The change occurred as dairying became more large-scale and market-orientated. In each case the change happened as a response to changing agricultural markets. The state played a key role in facilitating the change and encouraging men to assume responsibility. Some women were glad to be rid of such arduous work, others resisted the transition, albeit with limited success. There are three key sociological questions that arise from a review of the transformation of dairying: what was the role of the state, how adequate is patriarchy to explain the process, and why was the resistance of dairy women of limited impact? It is to these we now turn.

In the cases of Ireland, Denmark and Canada, we see clear evidence of the state's active role in promoting dairying as a male activity. Training, particularly for senior posts, was provided for men, and the nature of dairying was presented as having changed. Now it was scientific, whereas previously it had been based on women's practical expertise. As Cohen (1982) points out, the behaviour of the state was rational in terms of self-interest, and reflected women's economic position. The importance of dairying had grown, and if the industry was to expand accordingly, then men had to be involved in the process. They owned the land and controlled capital expenditure. The dominant ideology of the time further restricted women's access to public space, where agricultural business was transacted. However, it is unclear to what extent ideology was key in the transformation. Equally it could be argued that the state leaned on, and reinforced this ideology, in order to legitimise a course of action that was economically advantageous. Theorists of the state have argued that one way in which the state legitimates its own preferences is by reinforcing societal groups who support their preferences (Nordlinger, 1981). The state's invocation of the dominant ideology clearly legitimated a course of action that moved dairying to the male domain; it was too harsh, and inappropriate for women, and it was undesirable to have women occupying positions of prestige in public spaces. However in times of crisis, the state borrowed and leaned on other ideologies that facilitated women's dairy and field work. During the First World War, women in Ireland took up churning again, and women in the United States did enormous amounts of field work. In Australia, women received training during both world wars, although it was clearly presented as a temporary measure (Bell and Pandy, 1990). The interesting question is the extent to which the prevalent ideology was tied up with, and used to further, state policies.

In many respects, the transformation of the dairying industry represents a classic patriarchal process. Men appropriated a lucrative component of women's sphere of work, and men and a male state forced women out. At one level this is what happened. As a form of explanation, however, it is hardly adequate. It is limited in so far as it presents the activity as one of men asserting power over women (Pollert, 1996). For a fuller understanding, it is necessary to place the transition in the context of general history, of economic change and changing agricultural markets, notably the transition of dairying to a capitalist enterprise. The state did engage with men, but not solely in order to assist them exert greater control over women. The decision to do so was rational, in so far as it meant it promoted those actors who had resources.

Women were not passive non-actors in the whole transformation. Some were glad to be rid of the drudgery of dairy work. Dairying was not the only aspect of dairy women's lives. However some women did resist the loss of their occupation. This took a number of forms: letters to agricultural journals, organising through auxiliary groups of men's farming organisations, and antagonism towards male dairy managers. Their resistance was of limited success.

Hansen argues that rural women did not appear as a collective work force in history to the same extent as urban women because they did not have a common workplace from which to organise (1982, p. 226). Similarly, Davidoff (1974) suggests that Marx's description of French peasants as lacking a common identity of interests, and consequently lacking political organisation, was equally applicable to the isolation of wives. Mann (1986a) makes the same point about individual households being a hindrance to the organisation of women. To some extent this is true. Letters to agricultural journals are an individual act, and women's involvement in auxiliary groups were as part of a farm family. This line of argument, however, only takes us so far. Farm men were equally isolated in individual work units, yet agrarian organisation and protest was prevalent in the nineteenth and early twentieth centuries. More convincing reasons are women's limited access to public space, their primary identity as a member of a family rather than as individuals, and, most importantly, their lack of independent resources. Women's access to public space in farming organisations was as members of farm families. It was on this basis that women joined the American Grange and Alliance organisations. It is extremely difficult, almost unthinkable, in this context to express individual desires that may be contrary to the good of the farm, to which women also subscribed. Furthermore, to borrow Michael Mann's (1986) terminology, they were organisationally outflanked. They did not control resources and any efforts to maintain control of the industry were lame as a consequence. Skill alone did not provide them with the means to invest in the industry. The potential returns from engaging men in dairying were too important to the state to give priority to the wishes of women.

To summarise: the way in which the commercialisation of dairying affected women reflected the fact that they did not own the resources of the industry. While it was recognised that dairy work was carried out by women, there was little point in the state politically aligning with women to promote the expansion of the industry because women did not have the power to make decisions about investment on the farm. While some women protested at the changes that took place, they were organisationally outflanked; they did not have the organisational resources to successfully obstruct the proposed changes. The prevalent gender ideology of the time legitimated the state's course of action. The historical changes that happened in the dairy industry provide a useful comparative point with the organisational resources of the Canadian Farm Women's Network, which I will examine in Chapter 6, and with the current impact of social-democratic gender ideology on the behaviour of the Norwegian state, which I will consider in Chapter 8. In this and subsequent chapters, we see women's farming lives being shaped by a combination of state action from above, and women's organisation from below.

# 6 Women and Farming Organisations

## INTRODUCTION

When looking at inheritance and patterns of land transfer in Ireland throughout the twentieth century, we see that the patriarchal power of the heir within the family and the local community has diminished. None the less, the economic power, status and political power associated with land ownership has not. Land continues to be passed from father to son, and the various facets of power related to property ownership continue to be vested in men. This is evident when farming organisations are examined. The property owner is seen as the active farmer and this is reflected in the composition and agendas of farming organisations. They are predominantly male, and deal with a male understanding of farming.

In this chapter, I will look at how women are treated *within* farming organisations, and I will also look at the interaction *between* (male) farming organisations and women's farming organisations. In order to consider the former, how women are treated within farming organisations, I will return to Weber and his theory of bureaucracy. Weber suggested that within bureaucratic organisations, roles are occupied on the basis of merit and skill rather than inherited status. However, when I turn to look at women in farming organisations, the feminist critique of Weber's theory is more appropriate. Farming organisations are not gender neutral, and interaction with the occupants of roles is heavily influenced by the gender of the occupant. Symbolic interactionists focus on the construction of roles within organisations, and we see this happening on the basis of gender within farming organisations. Farming organisations are also engaging in the type of image-building exercises described by institutionalist theorists such as Meyer (1978). It is important for organisations to maintain their legitimacy in their wider environment. For farming organisations this means that as gender ideology changes, they must try to keep in step. As all-male organisations become increasingly unacceptable, farming organisations make certain attempts to include women and as I will describe, there is a growing incidence of the inclusion of women through sub-committees.

When I examine the interaction between male farming organisations and women's organisations, it is clear that both are an expression of the collective or created power discussed in Chapter 2. It is the effective organisation of farmers, although almost exclusively male, that gives them such political force. Women's organisations are equally a representation of collective power, and have brought about real change. However, (male) organisations are identified as farming organisations, while women's farming organisations are clearly identified on the basis of gender. When farming organisations are represented politically, it is male organisations that represent farming. Women's organisations serve to illustrate how farming agendas are set, and particular types of issues do not feature on that agenda. The women's organisations are not as included in the political apparatus as farming organisations. They are 'organisationally outflanked'.

The first part of the chapter considers the study of organisations, developments in the study of gender and organisations, and ends by outlining the key elements essential to the study of women and farming organisations. The second part of the chapter focuses on women within farming organisations. I look at those unusual women in farming organisations and sub-committees of women in predominantly male farming organisations. I also consider specific women's farming organisations, and the relationship between these and (male) farming organisations.

## STUDYING ORGANISATIONS

### Weber and His Critics

In their comprehensive overview of the study of organisations, Aldrich and Marsden (1988) suggest that such study is relatively recent, and has only flourished since the 1960s. However, Max Weber, writing at the turn of the century, is generally accepted as the founding father of the sociology of organisations. Weber argued that bureaucratic organisations were the superior form of organisation, and essential to the operation of large-scale industrial societies. In brief, bureaucratic organisations, as defined by Weber, are hierarchic and each role is clearly defined, with a fixed area of responsibility. People are appointed to their positions on the basis of merit and skill, which Weber saw as a significant improvement on despotic appointments and inherited status. Bureaucratic organisation is underpinned by rational–legal authority, or the acceptance of a set of impersonal rules. The legitimacy of this organisational form is based on the general acceptance of these rules. Inevitably, bureaucracy is accompanied by mass democracy, as signified by the demand for equality and fairness, rather than appointment by status or privilege (Weber, 1978). Weber identifies many positive features of bureaucratic organisation. It is transparent, and its roles are occupied on the basis of merit. Power and influence is restricted to those roles. It is organised on legal–rational authority, and any individual can refer to a book of rules if treated unfairly. Weber also had some concerns about the bureaucratic form of organisation that he described; we are all familiar with his reservations about what he called the 'iron cage' of bureaucracy, which he feared would dampen creativity and the human spirit. Of more interest to us in this chapter is his concern about the concentration of power in the hands of those who control the bureaucratic apparatus. While Weber believed in the technical superiority of bureaucratic organisation over all other forms, he remained pessimistic about the consequences for human freedom and happiness.

One of the interesting features of the study of organisations is the fact that, despite Weber's seminal position in sociology, for a long time it did not occupy a central position in the discipline. Instead it was confined to the almost exclusive remit of business studies. The main reason for this seems to be the nature of the debate that developed around Weber's work; the focus was on Weber's assertion that bureaucracies and formalised rules represent a superior form of organisation. The debate then centred on how to manage organisations most efficiently, how to ensure success, and other questions of most relevance to business organisations. Sociological questions of power, authority, inclusion, democracy and liberty, were less of a focus. Had the latter questions been addressed, it would have been more likely that an analysis of gender and organisations would have come to the fore sooner. While the work of Merton (1968), Blau (1963), Gouldner (1954), and Burns and Stalker (1966), all took on and critiqued Weber's theory of organisations from very different angles, their work shares a concern with the same central question of bureaucratic efficiency. Questions about the links between merit, skill and position in organisations on the one hand, and gender on the other, were much slower coming to the fore. If we look at farming organisations, we either have to accept that there are no women who are capable of filling organisational roles, or else we have to explore the possibility that a gendered social dynamic is interfering with the type of bureaucratic organisation outlined by Weber.

Before we turn to the contribution of gender and political studies of power in organisations, there are a number of facets of the more mainstream debate that are useful when we turn to look at women in farming organisations. Of particular interest is the question of legitimacy, and how central it is for an organisation's success to link its normative system with wider social values. Those who followed this line of enquiry have become known as 'institutional theorists'.

**Institutional Theory**

Weber argued that members of organisations attempt to establish and cultivate belief in their legitimacy. In other words, organisations attempt to have their procedures and actions deemed appropriate by prevalent social norms (Scott, 1992, p. 305). This formed the basis of theoretical developments in the 1970s, which moved away from a focus on the internal operations of the organisation to consider how the organisation interacted with its environment. Generally, discussion of the organisation's 'environment' refers to business elements, such as resources, financial support or information. Some institutional theorists have also examined how organisations derive legitimacy, power and authority from their status in the social environment. It is argued that the environment is more important to the definition of social validity than internal technical efficacy (Meyer, 1978). Institutional theorists argue that organisations adopt structures and procedures that are valued by the wider society. Thus an organisation may adopt certain practices or forms, not because they are thought likely to increase the efficiency of the organisation in executing its primary tasks, but because they are thought likely to enhance legitimacy in the eyes of significant environmental agencies (Meyer and Rowan, 1977; Bryman, 1993). Adept political work outside the organisation is as important as instrumental work within it to the long-term survival of the organisation.

Institutional theorists have been criticised for over-emphasising the impact of the external environment, for their tendency to view the organisation as a passive reactor to the environment, and for writing actors and agency out of an understanding of the behaviour of organisations (Bryman, 1993, pp. 86, 87). It is essential to maintain a perspective on senior decision-makers within organisations, and the power they wield. The 'dominant coalition', as Child (1972) christened it, exercises strategic choice, and decides on strategic action. However, while it is important to temper the institutionalist perspective, its validity must also be recognised. We only have to think of changed Nestlé policy following the widescale boycott of its products as a result of their promotion of powdered baby-milk in developing countries, to appreciate the importance of maintaining legitimacy to organisations. More recently, we have seen a restructuring of British and some Irish political parties in an attempt to increase the number of women representatives, a strategy that was developed in order to maintain legitimacy given the increasingly accepted belief in the importance of gender equality. This is the same environment in which farming organisations operate. In a wider social context where gender inequality is no longer acceptable, the persistence of all-male organisations is perceived as anomalous. Later in the chapter, I will argue that it is an attempt to maintain legitimacy and show some conformity to wider social norms that has prompted farming organisations to incorporate women, although the inclusion of women is usually through a sub-committee, or by the appointment of high profile, token women. None the less, the institutionalist perspective is important in helping us understand the relationship between women and farming organisations.

**Interactionists and Organisations**

While institutional theorists argue that organisational members' consciousness of the organisation's external image shapes behaviour, symbolic interactionists argue that it is internal interaction that shapes self-images and behaviour in the organisation. Goffman's (1968) work on asylums explored the way in which interaction between actors in an organisation shapes self-concepts and views of self. Interactionists argue that traditional studies of organisations try to understand human action by seeing actors' behaviour as if they act solely with reference to the normative demands made of persons occupying their positions in a social system (Manning, 1973). They argue that position is taken to explain behaviour, and the individuality of people in these positions is given insufficient attention. The important contribution that interactionists made to the study of organisations was to focus attention on how interaction shapes understanding of roles, self-perceptions, and an understanding of positions. This approach has rightly been criticised for its concentration on small-scale interactive contexts which ignores wider society. In addition it has been criticised for failing to give sufficient weight to objective restraints on social action. While acknowledging these criticisms, symbolic interactionism's focus on the interaction between actors in various roles is useful when we turn to women in farming organisations. The way in which they are treated by the media and other members of the organisation is shaped by the fact that they are women occupying a particular role.

**Gender and Organisations**

The feminist study of organisations may be considered part of the continuing debate about Weber's theory of organisations. In many respects, feminist work mirrors the work of institutionalist theorists and symbolic interactionists, but with a specific focus on gender. Feminist research has illustrated the importance of gender in shaping organisational roles, and the way in which gender is woven into the very fabric of bureaucratic hierarchies and authority relations (Witz and Savage, 1992). A gendered analysis shifts the focus from an instrumental one to a focus on the gendered nature of the implicit, informal ordering of organisations.

Feminist work has strongly argued that organisations are *not* gender neutral. It is not the case that gender differences are brought to neutral organisations. Rather, organisations are imbued with gender relations, and gender roles and identities are reinforced and created within organisations (Witz and Savage, 1992; Davies, 1992; Acker, 1991). Acker argues that one reason why organisations are theorised as gender neutral is because gender is difficult to see when only the masculine is present (1991, p. 163). This point will prove to be very important when we turn to look at women in farming organisations.

There are public organisations in which women occupy junior roles, or are completely absent. For example, Witz and Savage declare that most women in state bureaucracies are strangers in a male world. The same could be said of women in farming organisations. The different historical experience of men's and women's participation in particular organisations will affect and colour the way women enter the arena. Male homosociability, that is, the socialisation of men together, represents one way in which the management level of an organisation becomes a closed, gendered circle (Witz and Savage). In other words, if women do enter a male organisation, or even a senior position in it, social relationships reinforce gender differences, reflecting the argument made by interactionists.

One important question that is raised by studies of gender and organisations is whether the insertion of women into higher echelons of bureaucratic organisations will fundamentally transform these structures. Is it that organisations are staffed predominantly by men, or is it that they represent the collective interests of men? Is the issue that there are too few individual women in higher offices in organisations, or is it that the interests of women as a group are not represented? It is generally agreed that it is not enough to recruit individual women to senior bureaucratic positions. The collective interests of women have to be forged and explicitly articulated. This is a pertinent question for farm women. Would it be sufficient to have individual women recruited to the hierarchies of farming organisations? Can their collective interests be represented in farming organisations in their current form? Do women need to organise separately? Related to these questions, Witz and Savage classify three forms of state-related interventions that can lead to change in the position of women in organisations. Firstly, women may work *in* the state, that is as individuals or through the creation of women's units or departments. Secondly, women may work *through* the state, that is, use legal reform and participation. Thirdly, women may operate *in spite of* the state, that is, through the formation of alternate structures. Again, we will see that in various places, each of these strategies has been practised by farm women.

Acker identifies a number of interactive processes through which gendering occurs. While in practice they are part of the same reality, she argues they are analytically distinct, and when they are actualised, they give a clear perspective on the gendering of organisations. Firstly, she points to divisions along gender lines in organisations: divisions of labour, of allowed behaviours, of power, of locations in physical space. Such divisions in organisations are immediately obvious to casual observers. Secondly, she refers to the construction of symbols and images that explain, express and reinforce these divisions. They are found in language, ideology, popular and high culture, dress and the media. She gives an example of the image of a top manager, often portrayed as a confident man in the quintessentially masculine suit. Thirdly, Acker points out how interactions between men and women in organisations produce gendered social structures. All of these produce gendered components of individual identity within the organisation (Acker, 1991, p. 167). One further interesting component of divisions between men's and women's work relates to the clear value judgements involved in assessing the level of skill each entails. Frequently, women's work is deemed less skilled, or less taxing. This very much comes to the fore when we consider the relative value attributed to the work of women's and men's farming organisations.

So far, all the work we have examined on organisations is concerned with the *individual* organisation. Feminist and symbolic interactionists both take as their focus the individual organisation. The value of this work in helping us understand the position of women within farming organisations will soon become evident. However it does not take us far enough. To look at women and farming organisations, we need to be able to look at inter-organisational dynamics, and how the presence of long-established and well-resourced organisations impact on the formation of other organisations. Political science provides the most useful contributions to the study of this question, and it is to a brief review of this area that we now turn.

## Inter-organisational Dynamics

The contribution of Bachrach and Baratz (1962) to the study of power has already been considered in Chapter 2. One of the central thrusts of their work is that they move away from a behavioural and individualistic view of power. They argue that organisations constitute a form of collective agency, and that there is no reason to make this a second-rate form of agency compared to that of the problematic human agent (Clegg, 1989, p. 188). Bachrach and Baratz argue that many studies of power begin by studying the issues, rather than the values and biases built into the political system or organisations. Organisation is the mobilisation of bias, and some issues are organised in, while others are organised out. The mobilisation of bias becomes visible through expression of grievances or alternative forms of organisation outside, and in conflict with, the dominant form. In order to investigate the mobilisation of bias, it is necessary to begin by asking what are the dominant values, myths, established procedures and rules of the game.

Bachrach and Baratz introduce the notion of inter-organisational conflict. They argue that given the resources of highly developed organisations, the organisation of grievances may be inchoate and poorly resourced. I have looked at Michael Mann's development of this idea and his concept of 'organisational outflanking'. Mann argues that where people co-operate to enhance their collective power, social organisation emerges. While the organisation allows the pursuit of collective goals, a hierarchical form of power, which he calls distributive, develops within the organisation. Power is not shared equally and the 'masses at the bottom' have less power than 'the few at the top' (1986, p. 7). The power of those at the top to maintain their form of collective organisation is enhanced by two factors. Firstly, although anyone can refuse to participate, opportunities are usually lacking for the establishment of alternative machinery for the implementation of their goals. Secondly, the power of the existing organisation is strengthened if its control is institutionalised in the laws and norms of the social group in which it operates. Mann further argues, with respect to social stratification, that this is the fundamental reason why revolution has not occurred; the masses comply because they lack collective organisation to do otherwise. Because they are embedded within collective and distributive power organisations controlled by others, 'they are organisationally outflanked' (p. 8).

Mann and Bachrach and Baratz confront the question of inter-organisational dynamics. The existence, legitimacy and resources of an organisation impact on the development of other organisational forms. The ability to raise issues within an organisation, or the desire to form an expression of resistance requires resources, knowledge, and the ability and means to question the dominant rules of the game. The study of women and farming organisations should not focus only on internal dynamics, it must also consider inter-organisational dynamics, and this is most usefully done by following the lines developed by Bachrach and Baratz and Mann.

In the second half of this chapter, I want to use the work of Weber and his critics, symbolic interactionists and institutionalists, to examine how women are treated within farming organisations. The work of Mann and Bachrach and Baratz is useful when I look at the interaction between (male) farming organisations and women's farming organisations. Firstly, I will look at the case of the unusual woman who secures a position of authority in a farming organisation, using examples from the South of Ireland and Norway. Secondly, I will look at specific organisations of farm women, examining a Canadian example. Thirdly, I will examine the way in which some farming organisations create sub-committees where women participate, focusing once again on examples in the North and South of Ireland.

## WOMEN IN FARMING ORGANISATIONS

There is no country where women are well-represented in farming organisations. Indeed, farming organisations could accurately be called men's organisations. Take for example the South of Ireland. There are four women and 72 men on the national committee of the main farming organisation, the Irish Farmer's Association (IFA), and no men on the IFA's family-farm committee. There is one woman and 57 men on the national council of the Irish Creamery Milk Suppliers' Association (ICMSA). There is one woman on the executive of the Irish Co-operatives' Society (ICOS). The National Ploughing Association had until recently the unusual distinction of being the only farming organisation in Europe with a woman president, although recently the Norwegian Farmers' Union elected a woman president. In the North of Ireland the representation of women in farming organisations is similar, and it was only in 1996 that the Ulster Farmer's Union (UFU) appointed a woman to their executive committee of 80. What I want to look at now is how farming organisations treat the issue of gender, and how the media and male colleagues treat those few women who are members of farming organisations. Let us begin by examining the two unusual women who occupy the role of president.

### The Irish National Ploughing Association

The woman president of the Irish National Ploughing Association (NPA), Anna-May McHugh, was the secretary of the Association from 1959 until 1972. The founder president died in 1972, and she was appointed president in 1973. There was some dispute about the candidates for the post, and her appointment came as a total shock to her (interview, 1988). She said: 'I was shaking in my boots ... seven members (all men) met the night before the AGM in 1973 and decided they were going to get me elected.' Despite her most influential position, the organisation remains almost entirely male-dominated, although in 1997 her daughter was appointed development officer for the organisation. In 1997, the organising committee for the annual ploughing championships, hired in annually specifically for this purpose, was entirely female. The media continually focuses on these women in a way which emphasises their gender. A provincial newspaper featured the president with the following title: 'Anna is a lady in a man's world' (*The Nationalist*, 1989). The article went on to report: 'they say it is a man's world … but not down Ballylinan way. Anna-May McHugh holds one of the most important posts in the country, one usually associated with men.' In 1997, a feature on the organising committee was titled 'The women who run the ploughing' (*The Farmers' Journal*, *Journal Plus*, September 1997). In the same edition, a feature on the woman development officer was titled 'Ploughing her own furrow'. In each case, the media focused on the gender of the women, and in each case the features are in the *Journal Plus*, a gendered supplement to *The Farmers' Journal*, which we will consider in more detail below. Articles in the main body of the newspaper do not focus on men because of their gender, but rather on men as occupants of roles in farming organisations.

One feature of the annual ploughing championships is the farmerette class, the class for female competitors. There were over 300 competitors in the general ploughing championships in 1997, two of whom were women, and there were seventeen competitors in the farmerette class. The farmerette class was introduced in 1954, amid controversy and opposition from council members. Back then, the winner received a silver crown, a specially designed evening dress, and if she were unmarried and under 25 years, she would be presented with a cheque for £100 on the morning of her wedding, if she married before her 25th birthday. In recent years, journalists in the *Journal Plus* have started to ask if the title is sexist. Both the president and her daughter, a regular competitor in the class, defend the title (*Journal Plus*, September 1997). They argue that the title refers to the class, not the competitors, and in defence point out that the winner is called the Queen of the Plough, and not a farmerette, which it has been suggested is a demeaning term.

The media response to a woman president of the NPA illustrates how formalised organisational roles are imbued and shaped by the gender of the person occupying that role. While the gender of a male president of farming organisations is not a feature of media reports, it is central to reports of a woman who holds this position. Media reports constantly highlight the anomaly of a woman occupying such a role. Within the NPA, a woman president has not transformed the gendered nature of the organisation. Reports of the ploughing championships, and newspaper supplements about the event, feature men, and advertisements and articles also feature men. While this is ostensibly done in a gender-neutral way – the gender of the men is not the focus of the articles – the tacit understanding is that ploughing is a man's affair except for those separate activities specifically organised for women.

The processes identified by Acker (1991), through which organisations are gendered, are clearly discernible. Participation in the championships is divided along gender lines, with the much smaller participation of women largely confined to the farmerette class, itself classified on the basis of gender. The symbols and images that explain, express and reinforce these divisions are found in the ways both the popular media and the farming organisation deal with the female winner of her class. She is crowned Queen of the Plough, a symbolic gesture which once again focuses on the gendered nature of the winner. A queen can only be a woman, and the crowning of the woman at the annual dance in her honour reinforces the novelty and entertainment value of her win. Male winners receive trophies, gender-neutral prizes. Moreover, media reports of the ploughing do not focus on male participants as men in the same way they do with women participants. Both the treatment of women competitors, and media reportage of women within the organisation, constantly focus on and reinforce their gender.

The roles within the NPA are not gender-neutral as Weber's thesis suggest, but rather they are, as his critics argue, imbued with gender. Interaction with the occupants of organisational roles is shaped by their gender. The treatment of women within the NPA reinforces the idea that men are farmers, while women are more of a novelty.

### Norwegian Farmers' Union[1]

In 1997, the Norwegian Farmers' Union elected a woman president for the first time in its history. There were two candidates for the post, a man and a woman. A detailed account of the election process, and its reportage in the media, was undertaken by Verstad (1998), whose account of the Norwegian presidential election presents many of the same gendered issues that are illustrated by an analysis of the presidency of the Irish NPA.

Verstad argues, in a striking parallel to the case of the Irish NPA president, that it was the election committee that was the driving force in the election of a woman as president of the Norwegian Farmers' Union (NFU), despite signals from the grass roots. Within the election committee, it was the organisations who represent rural women and rural youth who argued the case for a woman president. Verstad describes the traditional perception in the NFU and the media that women's participation is gendered, without registering any comparable belief that men's activities in NFU are gendered. The arguments put forward for the candidates presented the man in a purely action-oriented way, while the woman candidate was given a symbolic function which emphasised the fact that she is a woman. For example, the woman candidate was asked what she as a woman could do for the organisation, while the male candidate was not asked what he as a man could do for the NFU (Verstad, p. 18). One of the national daily papers ran an article on the elections before the candidates had been nominated, under the headline 'We Want a Woman in the Director's Chair'. The headline reinforced the gender of the woman candidate, rather than her ability to successfully perform the functions of the presidency.

Verstad notes that it is hoped the woman president will change the role of women in Norwegian agriculture, leading to greater equality. The president will serve as a role model for younger women, and she will encourage women to use their allodial rights. It is too soon to assess the impact of a woman president of the NFU. In itself, her election is important and symbolises change. However, as in the Irish case, media reports of her role continue to mark her work and position on the basis of her gender, and by so doing stress the irregularity of a woman holding this role within the organisation.

## Exceptional Women

Media accounts of women who do occupy roles in farming organisations frequently present them as exceptional women, highly motivated, talented, and high achievers. Some Australian researchers have called this the construction of women as heroines and saviours (Grace and Lennie, 1997). One member of the Irish Farmers' Association (IFA) that I spoke to about the low attendance of women at local meeting said, 'there would be very few women at these meetings, okay … but while they might be few in number, any of them that would be there would have very strong views on things and would be well able to speak up for themselves' (interview, 1989).

Recently in the North of Ireland, the first woman was elected to the Ulster Farmers' Union executive committee. Of the 80 members on the committee, she is the only woman. A recent article featured her under the headline 'A Strong Voice for Northern Women' (*Journal Plus*, 20 September 1997). The article begins by describing her as 'blessed among men. She is the only woman on the 80 strong Ulster Farmers' Union executive'. It details how in 1996 she received an MBE for her services to rural women, has worked with refugees in Yugoslavia, and has made 'hair-raising journeys in Columbia'. She is the deputy president of the Associated Country Women of the World, and a candidate for the forthcoming presidential elections of the organisation. The long article ends by saying that there is 'a thread which is now drawing farm women the world over away from the sideline into the maelstrom of policy making and decision taking'.

While there is no doubt that women who do participate in farming organisations are successful, talented people, the emphasis given to this by the media and by men in farming organisations suggests that it is women who must rise to the challenge to participate. Ironically, it is as if the absence of women from farming organisations is seen as lack of interest, or lesser commitment to farming. Those exceptional, strong-willed women prove that women can participate, so why do more women not follow their example? It legitimises and perpetuates the male dominance of agricultural organisations by suggesting that it is men who are generally more capable, more interested in farming, and have views and opinions they are prepared to articulate. Attention is diverted away from the structure of farming organisations, which subtly excludes women. It is particularly important to stress the way in which the actions of farming organisation are gendered, even though they are presented as gender-neutral. One agricultural adviser I interviewed, said 'there is nothing stopping any woman doing a course here or from getting more involved in farming matters. She just has to be brave enough and have neck enough. She must educate herself about it too. But it is up to her to go and do it.' Similarly, the Ulster Farmers' Union president – the Union we have already noted has one woman on its executive committee of 80 and has no committee chaired by a woman – has stressed that there are no obstacles to female involvement in the union (*Journal Plus*, 16 March 1996). These comments focus attention on the abilities of individual women rather than the practice of organisations. The women I interviewed in the South of Ireland in the late 1980s, and the women in the North of Ireland in the mid-1990s, all reported how notification of farming courses and training sessions are addressed to their husbands. The tacit understanding here is that training is for men. Moreover, they all had a strong impression of Farmers' Unions as being the concern of their husbands. One woman told me that it is men who are involved in the Irish Farmers' Association (IFA), and 'that is where they get down to the real business of farming'. In 1989, the IFA organised a symposium titled 'The Farm Family and The Law'. The publicity for the day stressed that it was open to all members of farm families, especially women. By emphasising that the invitation was inclusive of women, it was clear that the IFA did not expect them to attend otherwise. I attended the day, where a sizeable minority of women were present. Four of the five chairpersons were men, and all four speakers were men. The women present were welcomed at the beginning of the day, and they were thanked at the close of the day for attending. The attention their presence received reinforced how unusual the situation was. The fact that the IFA could prompt women to attend by specifically targeting women illustrates the gendered nature of their day-to-day policy.

Men, as a result, are more visible in farming organisations; it is men from these organisations who represent farming interests on state and semi-state bodies, and they appear more active, interested and competent than the absent, silent women. In a tautological fashion, their predominance in farming organisations is considered legitimate, and farming is presented as a male occupation. The stress placed on the occupation of roles by unusual women because of their 'exceptional' merit and skill, detracts attention from the organisational culture which tacitly excludes many women.

**The Farming Media**

Recently, research, and in particular discourse and textual analysis, has illustrated the way in which the farming press reinforces the gendered nature of farming, as many of the examples used in this chapter illustrate. Norwegian research has illustrated how the farming media focuses on women's gender (Brandth and Haugen, 1997; Verstad, 1997), and Irish research has considered the main national farming newspaper, *The Farmers' Journal* (Duggan, 1987; Shortall, 1992). I want to turn our attention now to the structure of this newspaper.

*The Farmers' Journal* (hereafter referred to as *The Journal*) is the farming newspaper with the widest circulation in Ireland, having a Northern and Southern edition. It is published weekly, and has been extremely popular since the 1940s. *The Journal* is divided into two parts, the main body of the newspaper, and a pull-out magazine, called *Journal Plus*, usually about sixteen pages. Most of the articles written in the main part of the newspaper are written by and feature men. These include articles on contemporary farming matters, containing information on current farming affairs, fertilisers, farming methods, equipment, financial matters, farming advice, and a letters section. Advertisements are for farm equipment, fertilisers, farm insurance, farm overalls, and again almost exclusively feature men. The pull-out magazine has an entirely different focus. There are considerably more female journalists, the editor is a woman, and the articles are not only about farming, but also about country life, and there are articles on fashion, cookery, gardening, films, health issues, and child care. The letters to the editor are more usually about current affairs and issues of particular concern to women than farming. Advertisements are for household goods and appliances, fashion and knitting wool.

The *Journal Plus* has been important in providing a forum to high-light women in farming, and in providing a space in *The Journal* where women can articulate their views. At the same time, *The Journal* is clearly organised along gender lines: the 'serious' farming issues, of interest to and discussed by men, are presented in the main body of *The Journal*, and the less productivist-orientated issues, concerning the family, health and safety on the farm, the role of women in farming, and other general issues of interest to women, are presented in the pull-out supplement. Leaving aside for the moment the questionable view that health, safety and family issues are not central to the success and continuation of farming, the organ-isation of *The Journal* rigidly represents and reinforces gender divi-sions within farming. A feature on a woman mart manager, or the woman president of the National Ploughing Association, is never carried in the main body of *The Journal*. The image presented is that gender (read women's) issues are dealt with in the *Journal Plus,* while the main body of the newspaper is gender-neutral. It is as Acker (1991) notes, an instance of where gender is difficult to see because only the masculine is present. *The Journal* is another example of an organisation that invents, reproduces and disseminates a gendered understanding of farming.

## WOMEN AND FARMING COMMITTEES

Within farming organisations, there is often a committee, comprised mainly of women, which has a brief to deal with issues relating to the farm family, or/and farm women's issues. For example, The Irish Farmers' Association (IFA) has a Farm Family Committee, which was established in 1976. The Committee's remit is to look after issues of relevance to farm families and farm women. There are no men on the Farm Family Committee. Similarly, the Ulster Farmers'

Union – (UFU) established a farm-family committee in 1996, and of the 28 members, 23 are women. The issues that farm-family committees address are quite similar to those of the Canadian Farm Women's Network. The UFU farm-family committee, for example, identified stress as an issue in urgent need of attention. The BSE crisis, farmers' isolation, long working hours, and the bleak future of farming have made it a very stressful occupation, with an increasing rate of suicide. The UFU farm-family committee designed a rural stress initiative card, which they circulated in doctors' surgeries, agricultural offices, and marts, highlighting existing helplines to assist people in distress, such as the Samaritans and the Citizens' Advice Bureau. As well as looking at the role and work of women, the farm-family committee also has an extensive brief which includes issues such as the promotion of food produce, profiling the benefits of keeping rural communities vibrant and economically viable, tackling issues such as rural isolation, and lobbying for the maintenance of the rural infrastructure.

While they do not formally constitute a farm committee, farm women's training groups are a new feature of the agricultural advisory service in Northern Ireland. Four of the 42 agricultural advisers in Northern Ireland are women. One of these felt that farm business plans would be improved if women were more vociferous in stating their opinions and ideas when farm business plans were discussed. She was also aware that women were doing farm accounts and would benefit from a more complete understanding of the financial state of the farm. Despite repeated requests for women to come along to existing training provisions, women did not participate. She decided to organise some training specifically for women, and discussed the idea with two of the other women advisers. Then they actively solicited women on the farms with which they had contact. These three women advisers are responsible for the creation of five farm-women's groups. These groups will be discussed in further detail in the next chapter. What is relevant here is that they are a specific, overtly gendered provision for women, colloquially known as the 'ladies' group' or the 'farm wives' club'. The provision is very patchy, limited to those areas where there are women advisers. Approximately 80 women participate in total. It is worth noting that the advisers formed the groups in response to a practical, technical problem. They thought that the farms with which they had contact would be more effective if the women on those farms were trained. In other words, the formation of the groups was not motivated by political, moral or feminist considerations, although clearly there is a gender factor involved in the fact that it was women advisers who recognised the lack of training for women as problematic.

Earlier, we looked at how the institutional theorists argue that organisations attempt to have their procedures and actions deemed appropriate by prevalent social norms. In a social context where gender equity is generally accepted as an appropriate moral objective, the existence of entirely male organisations are less acceptable. Given this climate, in order for farming organisations to maintain their social validity, it is important for them to adopt structures and procedures valued by the wider society. In this context, the presence of women becomes important, regardless of whether the organisation considers it likely that they will increase its efficiency in executing its primary tasks. From the organisation's point of view, farm-family committees are an important legitimacy claim. The same is true of the farm-women's groups within the agricultural service in the North of Ireland. In the last few years in Northern Ireland, forms of inequality and discrimination other than religious are receiving increased attention (McWilliams, 1993). Even though the women advisers have received limited organisational support for the formation of training groups for women, they are pointed to as one of the Department of Agriculture's provisions for women. This changing social milieu may provide a wider context of support for the farm-women's groups. Equally, they may persist in their current format, which is to provide women with some limited access, without questioning the gendered nature and structure of agriculture training.

The farm-family committees operate as autonomous units within farming organisations. The composition of the groups is seen as gendered, that is, consisting primarily of women. Clearly the composition of the rest of the organisation is also gendered, consisting primarily of men, but that is hidden because of its primary identity as a farming organisation. The issues that the committees address are also seen as gendered, those of concern to women, or of a caring nature. The dichotomy between farming and the farm family is created and reinforced within farming organisations. Scott (1992) asks what difference it makes if a group is admitted to an organisation. He argues that the difference should be reflected in the goals pursued by the organisation. Admission is an empty victory if the new partners cannot affect the definition of the goals to be served (p. 295). The committees of farming organisations on which women serve have not redefined the goals of the organisation. On the contrary, they almost reinforce gender divisions, while legitimising the organisation in the wider social context.

## FARM-WOMEN'S ORGANISATIONS

Having looked at the success of a few women who have managed to break into the higher echelons of farming organisations, it is clear that, on its own, it does not necessarily mean that the character of the organisation becomes less masculinist, or more feminised or woman-friendly. One alternative to the obstacles women face in mainstream farming organisations is the formation of farm-women's organisations. Farm-women's organisations have emerged in a number of countries, for example, Australia (Teather, 1996), the USA (Haney and Miller, 1991), and Canada (Shortall, 1993, 1994, 1997).

### The Canadian Farm Women's Network[2]

In this section I want to look at the Canadian Farm Women's Network (CFWN). It would be a misconstrual to suggest that women primarily organised as a response to (male) farming organisations. None the less, in the Canadian case women organised because there was no other forum where they could express their concerns and tackle the issues they wanted addressed. The CFWN provides an example of how certain issues get 'organised out', and how the rules of the game and the issues of concern in mainstream farming organisations are determined by those who participate. The existence of farm-women's organisations alongside farming organisations allows us explore the dynamics of inter-organisational power relationships.

*Formation of the organisation*
The development of farm-women's groups in Canada since the mid-1970s has been referred to as 'the new farm women's movement' (Harkin, 1991). It is important to distinguish the groups under discussion as *farm-women's groups*, that is, as groups of women who identify on the two criteria of womanhood and farming. There have been many and various women's groups in rural Canada for a long time. Farm women have been members of these, but they were not specifically farm-women's groups. Women for the Survival of Agriculture (WSA) was formed in 1975 in Winchester, Ontario. This was the first farm-women's group in the 'new farm women's movement'. In 1980,

Women in Support of Agriculture was formed in Prince Edward Island, and L'Association des Femmes Collaboratrices in Quebec. By 1991, there were 42 new farm-women's groups (Farm Women's Bureau, 1991). In 1985, it was decided to organise a national network which would co-ordinate the provincial networks and local groups. The Canadian Farm Women's Network (CFWN) was formed.

The nation-wide farm crisis is reported by farm women and academics as an important motivating factor in the organisation of farm-women's groups (Harkin, 1991; Bruners, 1985; Haley, 1988). Increased farm bankruptcies, stress related to financial difficulties, cutbacks in rural social services and other problems faced by the farm family became the issues around which farm women organised. They were, and are, committed to the survival of the family farm. Another contributing factor was the women's liberation movement. The movement meant that there was increased attention paid to the concerns and views of all women, including farm women, and women came to see themselves in a new light (Bruners, 1985). Furthermore, following International Women's Year in 1975, there was increased state funding provided for projects dealing with women's issues. In Canada, farm-women's groups broadened their mandate in order to qualify for such assistance (Harkin, 1991). Yet another important catalyst was the Murdoch case of 1973. Irene Murdoch had worked for twenty years of her married life on the Murdoch farm, maintained a home, supplemented the farm income with a market garden, and provided capital to purchase additional land and machinery. Upon divorce, the Supreme Court of Canada ruled she was only entitled to $200 a month from her husband, which even at that time was well below the poverty line. The Law Reform Commission of Canada and the Canadian Federation of Agriculture were among the groups who lobbied for reform of property law following this decision. The support of these powerful organisations was crucial, and at the time it increased interest in and awareness of the value of the unpaid contribution of women on farms to the farm business. Irene Murdoch returned to the courts and in late 1976 was granted $65,000, which represented one quarter of the Ranch's value (Bruners, 1985, pp. 18–19). Since the Murdoch case, the law has changed substantially and in all Canadian provinces wives are entitled to one half of all the family assets.

*Why organise?*

The issues that the farm-women's organisations have addressed concern farming, the farm family, and farm women. While the three areas are analytically distinct, in practice they are intertwined and part of the same reality. Let us consider each of them in turn.

*Farming*
The founder of WSA, the first of the farm-women's organisations, stated that the initial goal of the group was to deal primarily with agricultural matters (Harkin, 1991, p. 2). The CFWN has lobbied government at a provincial and more recently at the national level about tractor-licensing laws, the continuation of the Canadian Wheat Board, regulations regarding the freight of cattle, and GATT negotiations. Farm-women's groups have been interviewed about issues such as the effects of free trade on farming and interest-free loan programmes for farmers. More recently, they have given some attention to sustainable agriculture and the environment. One distinctive feature of farm-women's groups is that they are not organised around particular sectoral farming interests, as for example dairy producers, or grain producers.

*Farm family*
Farm women cite the implications of the current farm crisis for the farm family as another reason why they organised (Farm Women's Bureau, 1991). Through research and lobbying, farm-women's groups have drawn attention to the effects of economic crises, and the resultant financial difficulties, on the mental health of the family (for example, Women of Unifarm, 1979). In times of difficulty, shortcuts are taken, often with detrimental consequences for family members. The farm-women's organisations have highlighted the risks of undertaking child care alongside farm tasks, and frequently report statistics on child accidents on farms, with advice on how to avoid such misfortunes (see, for example, the Saskatchewan Women's Agricultural Network (SWAN) Newsletter, Spring 1997). In addition, they have lobbied government about rural child-care facilities. The women's organisations have also commissioned research on safety precautions regarding farm machinery and the storage of chemicals. They have distributed farm safety packages, and organised farm safety days (CFWN Newsletter, 1991:3, 6).

*Farm women*
The very existence of farm-women's organisation has increased the profile of women in what is traditionally a very male-dominated sphere. In addition to lobbying about farming and farm family issues,

they have also lobbied about issues that affect them as women on farms. The CFWN has considered the issue of women and ownership of land. An Ontario women's group published (1987) and later revised (1994) a manual titled, *Cover Your Assets: a Guide to Farm Partnerships*. This booklet details both the benefits for the farm (tax advantages, loan entitlements), and the necessity for women of legalising their stake in their business partnership. Women are encouraged to safeguard their legal position after entering a farm business.

Farm women quickly realised that they met few women in farming organisations, in government farming bodies and committees, or on marketing boards. When they raised this issue, they were often told no suitably qualified women existed. The CFWN decided to form a 'talent bank' of farm women; this is a directory of qualified farm women who can be proposed as vacancies arise in farming organisations. Provincial networks provide training for women who wish to sit on boards and councils. This strategy has had some limited success. None the less, women continue to struggle to ensure adequate representation (*Farm Living*, 9 November 1995). Farm women have noted, too, that it is more difficult to include women in traditional, grassroots farming organisations than it is in federal or provincial government farming organisations, boards and committees (*Farm Living*, 23 October 1994). The CFWN has called for changed regulations regarding membership, and has also urged farming organisations to deliberately recruit, support and educate women at grass-roots, district and regional levels.

The Canadian Farm Women's Network highlights the limited number of women in farming organisations. Through their talent bank and lobbying of organisations, they actively try to increase the representation of women. In doing so they are questioning both previously taken-for-granted farming norms, and the legitimacy of institutions that are unrepresentative on a gender basis.

The CFWN, and most provincial farm-women's networks, issue regular newsletters. This provides an alternative source of farming information for women. It also provides a forum where they can discuss and debate issues of particular relevance to women, such as partnership arrangements, the availability of credit for women operators, and the status of women's farm work. In addition, the farm-women's networks provide training for farm-women's groups to deal effectively with the media. Women have received training in public speaking and drafting press releases. In recent years, many provincial newspapers with farming supplements have carried regular articles about the farm-women's networks, and short pieces written by farm women.

The farm-women's newsletters have a wide circulation amongst members. Although modest, there is increased representation of women in more mainstream farming media. Again, the farm-women's movement has acted as a key force for bringing about change through providing an alternative form of communication, and through training women to effectively deal with the general farming media.

A key concern of the CFWN is how to define farm work and how to ascertain the value of farm-women's work. It is a frequently discussed topic in the newsletters and at farm-women's conferences. Since 1990, editorials in the CFWN Newsletter have urged women to make their farm work visible by changing the title of their work so that it can be counted in the census of agriculture. Newsletters feature articles by farm women that say discussion and dialogue has raised their appreciation of the often invisible and unrecognised work they do (OFWN Newsletter 3, 4, 1991). One of the most significant achievements of the farm-women's network was their successful lobbying of Statistics Canada to change the Census of Agriculture questionnaire. Prior to 1991, it was only possible to cite one person as farm operator. Women argued that in instances of partnerships between men and women, it was usually the man who was cited in the census form as the farm operator, thus rendering women invisible. This practice was changed in 1991, and now up to three operators can be named. It is generally accepted that the change came about because of effective lobbying by the farm-women's network (Farm Women's Bureau, 1996). The change only increases the visibility of women in terms of the typically 'male' farm work that they do, but it marked a very important triumph for the farm-women's network. It is an example of collective power, where by coming together, farm women have increased their power and enacted real change.

## Farm-Women's Organisations and Farming Organisations

The farm-women's organisations have enacted personal and social change. They sanction a public role for women in an industry where this has not traditionally been accommodated. In the process, through their discussions and lobbying, they are changing the way in which farm women perceive their role, and at some level how the general public perceive farm women. Participation in such organisations has led to greater consciousness amongst farm women who belong to them. For example, farm women in Prince Edward Island reflected at their tenth annual meeting on how their provincial group had given them a positive view of their occupation and of their role as farm women (CFWN Newsletter, 1991: 3, 6). The sense of empowerment and confidence is obvious in the newsletters. The consequences of active organisation have been enormous for farm women.

Yet the very title of this section embodies one of the issues I want to address: farm-women's organisations state the gender of participants in their title, while farming organisations do not. Regardless of the extent to which farm-women's organisations profile farm women, or extend the understanding of the full range of issues involved in family farming, farming organisations as they stand consist primarily of men, and are largely uni-focused on a production-orientated understanding of farming.

The farm-women's organisations illustrate the manner in which some issues are 'organised out', or as Crenson (1971) described it, become 'non-issues'. At present, farming organisations do not address issues relating to farm women, such as their legal status in the farm, statistical accounts of what constitutes farm work, health and safety issues on the farm, or matters relating to child care on farms. Using Bachrach and Baratz, the farm-women's organisations can be understood as exemplifying an expression of grievance, through their lobbying and organisational activities. They are outside mainstream organisations, and it is only by organising in a different forum that they have the means to address their issues and concerns.

There are other ways in which the farm-women's organisations are outflanked. They are not represented on state and semi-state government bodies in the same way as mainstream farming organisations. The latter continue to represent a uni-dimensional understanding of farming, while the farm-women's organisations, from outside the central arenas of power, lobby government and the farming organisations toward change. In an ironic way, part of the reason for this is that farming organisations are seen as representing farming issues, while farm-women's organisations are seen as primarily addressing gender issues. The source of each organisation's funding reflects this point. Farm organisations are funded through members' subscriptions, and in some cases by a levy on production. In other words, farming organisations are funded by farm producers and farm production. Conversely, the Department of the Secretary of State is one of the main sources of funding for the Canadian Farm Women's Network. The State Department considers it progressive for farm-women's groups to separate farm issues and farm-women's issues, and will fund activities relating to the latter sphere (see Dion, 1990, pp. 26–31). However, farm-women participants at a Canadian Federation of Agriculture Conference (April 1991, Ottawa) found these guidelines unhelpful. Funding can be obtained for the consciousness-raising of farm women, but not for the organisation of educational courses on farming matters. The women felt that farm-women's issues and farm issues are intertwined. For example, one concern of farm women is their lack of representation in farming organisations and on marketing boards. In order to address this, women must be knowledgeable on farm issues. Therefore, if the State Department will not fund farm-women's groups to organise farming courses, on the grounds that they are not of *specific* relevance to women or women's groups, they may in fact be unintentionally impeding the participation of women in the traditionally male farming arena. The funding guidelines fail to see women struggling with an occupation that does not recognise them, focusing on them as women.

The issues that the farm-women's organisations have addressed illustrate many of the processes by which organisations are gendered, as described by Acker (1991). Firstly, the division of space in farming organisations, or the lack of it for farm women, has meant they have had to create a new forum. Secondly, they have created their own newsletters as a way of circumventing their absence, and the absence of the issues they wish to discuss, from the general farming media. Thirdly, the existence of the farm-women's organisations, and their interaction with farming organisations, ironically recreates an understanding of gendered issues in farming as something the farm-women's organisations deal with, while the farming organisations deal with farming matters.

A complicating feature of the farm-women's organisations is their determined efforts to save the family farm and the farming industry, which in many respects is the source of their unequal position. American farm women, interviewed by Rosenfeld, disliked what they saw as the women's movement's disparagement of the traditional family and of women's role within it. These women saw themselves as part of the *family* enterprise (1985, p. 268). In a similar way, the primary concerns of the American farm women's group WIFE are the family farm and farm-family welfare rather than themselves as farm women (Haney and Miller, 1991). Australian farm women have similar problems with the idea of feminism (Alston, 1990; James, 1992). Hostility towards men is inconceivable in a system that depends on such a high degree of co-operation. Furthermore, farm women and their husbands are joined by marital ties and bonds of affection and do not want to become business adversaries. There is no doubt that farm women do have a unique relationship with their families. The family farm is, along with small family businesses, one of the last residues of the organisation of labour through the household. Feminism is a term that many farm women find problematic. One editorial in the Ontario Farm Women's Network Newsletter clearly outlined some of the barriers to participation faced by women, in agriculture. It stated: 'if women are successful in attaining high-powered positions, one of their responsibilities should be to support other women and "women's issues", a term I hate to use. There are no "women's issues", they should be called "family issues", or better yet "society's issues"' (OFWN Newsletter, 3, 4, 1991). The difficulty however, is that while women see their concerns as tied to the family farm, there is no move by farming organisations to actively embrace women.

CONCLUSION

Farming organisations are an example of collective power. However, they are generally an example of male collective power, as women do not participate to any great extent. It tends to be the property owner who is involved in farming organisations. The fact that property owners are almost always male has led to the gendered construction of farming organisations. The treatment of women within farming organisations serves to reinforce gender roles within farming organisations. Contrary to Weber's theory of bureaucracy, sexuality mediates and shapes formalised roles and organisational practice in farming organisations. We have looked at three instances of women's relationships with farming organisations in this chapter: exceptional women who hold positions in farming organisations, specific farm-women's organisations, and committees within organisations that consist mainly of women. Each case illustrates the gendered nature of farming organisations.

   Drawing on symbolic interactionism, we see how the rare presence of women is reified in how they are treated by the media and by farming organisations; 'lady in a man's world', 'farmerettes', and the 'Queen of the Plough' are all instances of the way in which women's gender is the focus of attention, rather than organisational practices which may be accountable for the low participation-rate of women. As Acker rightly points out, symbols, the media and images create and reinforce the gendering of organisations. The presence of a few women in the higher echelons of farming organisations is not enough to change the weight of organisational culture. Treating women as a novelty limits the extent to which they are taken seriously, and reinforces their anomalous position within the organisation. In turn, this limits their political power in farming, both within farming organisations and in their representation at state and semi-state levels.

The farm-family committees within farming organisations, on which women serve, are a clear example of the gendered division of space and tasks described by Acker (1991). The committees are confined units, and have little impact on organisational goals.

One of the difficulties faced by farm-women's organisations is that their existence suggests that women constitute a special interest group within farming. They are seen as having specific concerns and issues that they wish to advance. The same is true of farm-family committees within farming organisations. In short, the fundamental constitution of farming organisations, which has 'organised out' women and a broader range of farming issues, is not questioned or threatened. They persist unfettered alongside these new (and peripheral in their eyes) developments. The exclusion of women from farming organisations is part of the reason for a partial understanding of what farming entails, from business issues, to health and safety, to the stress and isolation of the occupation, and to the recognition of roles of farm family members. The difficulties of advancing a feminist case from within farming organisations is compounded by the bonds of affection and reciprocity that are central to family farming. As long as women are organised outside of the mainstream organisation, there will be conflict or division. As long as they are not fully incorporated within farming organisations, a real gendered division persists. The Canadian Farm Women's Network represents an expression of grievance at women's exclusion from existing channels of power and decision making. Their agendas illustrate how there are a whole group of people farming who have concerns that do not appear on standard farming agendas. Their position outside the established institutional apparatus reflects the extent to which they are organisationally outflanked; they do not have access to the same organisational resources as (male) farming organisations.

Whether we look at the impact it is possible for individual women to have within farming organisations, or the amount of change farm-women's organisations can achieve, or the inclusion of women in farm-family committees, the focus is always on women. It is farming organisations that need to be fundamentally reformed if farm women and the wider range of issues they bring to the table are to be treated seriously. A critical examination of what farming organisations currently represent illustrates that they present only a partial perspective on what constitutes farming, and a limited understanding of who are the players. The players they primarily represent are property owners.

# 7 Women and Agricultural Education

## INTRODUCTION

In its broadest sense, education is simply one aspect of socialisation. It involves the acquisition of knowledge and the learning of skills. It is also generally accepted that education, intentionally or unintentionally, helps to shape beliefs and moral values.

While the study of education is central to mainstream sociology, the same issues and questions have not been at the forefront of analyses of agricultural education and training. Agricultural education and training (terms to be used interchangeably here) is generally assessed in terms of efficiency and effectiveness, with little recourse to the sociology of education. In fact, agricultural education is influenced and biased by wider social factors, that have been analysed in the mainstream sociology of education. These include the effect of power, how education exacerbates or ameliorates gender inequalities, the ability of education to correct inequalities in the more general social milieu, and the social construction and different valuation of different types of knowledge. It is these questions, from the perspective of farm women and agricultural education, to which we will turn in this chapter.

The low participation-rate of women in agricultural education programmes has been noted, and various explanations and analyses have emerged. Largely these relate to the fact that men are property owners, and are seen to fill the occupational role, and hence undertake occupational training. Men know they will inherit the farm, and therefore it makes sense to undertake agricultural training (Delphy and Leonard, 1992; Shortall, 1996). We are back to property again. Women do not usually know they will work on a farm until they marry a man who has inherited a farm, so to undertake training prior to this would constitute an irrational choice. In other words, the means of entry to farming greatly shapes the relationship of men and women to training. Difficulties conceptualising the work women do on farms exacerbates the problem of training provision. If the work women do is not recognised, it is impossible to provide training. The provision of agricultural education for women is of pragmatic and equitable

115

importance. Pragmatically, education increases the efficiency of farm-women's work, while in terms of equity, providing training for women renders their work and work-roles visible and valued. The social construction of agricultural knowledge reflects a certain understanding of the work and role of women on farms.

I will begin this chapter by discussing the social construction of knowledge, and then consider the social construction of agricultural knowledge, and to whom it is considered appropriate to impart this knowledge. I will argue that it is fundamentally shaped by property rights. I will consider the curricula of education programmes and examine the taken-for-granted consensus that it adequately covers agricultural education and training needs. We will also look at how it reifies a particular understanding of what constitutes farming and who carries it out. The way in which knowledge is stratified will also be examined, and this will become pertinent when we move on to look at the education and training provisions for women that are currently being made available in the three study areas considered in this chapter: the North of Ireland, Canada, and the South of Ireland. In each case the provision is understood as a provision for 'women', and additional to the mainstream provision which continues unrevised. It is not understood as broadening the agricultural educational curricula.

## THE SOCIAL CONSTRUCTION OF KNOWLEDGE

Commonsense assumptions about knowledge usually involve the belief that it is objective, 'out there', and imparted through the educational system. However, scholars who have sought to discover 'objective' knowledge, have had to contend with the fact that the search for and discovery of knowledge is socially organised (Apple, 1979). In other words, knowledge is not so much a 'factual' or 'real' representation of the world, but more an outcome of attempts to interpret and apply some description of the world. Knowledge is produced by commonsense actors, hence it fundamentally lacks objectivity (Blum, 1981). There is nothing 'natural' or 'divinely ordained' about the knowledge that is presented through educational structures. The selection and organisation of knowledge involves conscious and unconscious social and ideological choices. Think for a moment about the religious instruction in schools: in the South of Ireland it is Catholic instruction, in Greece it is Greek Orthodox, in Saudi Arabia it is Muslim. Or consider the foreign languages we learn in schools: in most of Europe it is generally other continental European languages, while in East Germany, prior to the collapse of the Berlin Wall, the language learned in school was Russian. With both religious instruction and the languages we can choose to learn, conscious and unconscious social and ideological choices are apparent. We should not assume that curricular knowledge is neutral.

The key issue, then, is that rather than accepting the curriculum as given, we need to problematise it, so that its latent ideological content can be uncovered. Whose knowledge is it? Who selected it? Why is it organised and taught in this way? To which particular group? If these questions are not asked, we end up in a situation similar to that described by Bachrach and Baratz (1962), whereby there are non-decisions about the curricular content, because it is presented, and understood, as being beyond question (Young, 1981). All of these questions are pertinent to agricultural education.

## THE SOCIAL CONSTRUCTION OF AGRICULTURAL KNOWLEDGE

In order to understand the construction of agricultural knowledge, it is necessary to return to an analysis of the way farm labour is conceptualised. In general, it is productive farm labour that is recognised. Furthermore, the definition of productive labour is limited to labour associated with commercial agricultural production. Only remunerated work is considered 'real' work, and indeed, only commodity production is considered 'productive' (Sachs, 1983). By and large, it is the work of farm women that is omitted by this narrow definition of farm labour. Much of the work which women do in sustaining the family household and the farm business is not recognised, and rendered invisible (Reimer, 1986; Whatmore, 1991a, 1991b). The effect of the social construction of agricultural knowledge on training for women is twofold: firstly, much of the farm work that women do is not recognised as such and there is no relevant training available, and secondly, even if they do perform recognised farm work, they are often not given due recognition and included in training programmes.

Women have been seriously affected by the imposition of the business model on the family labour farm. They are rarely designated as farm operators or owners, very few are designated as paid labourers, and consequently, much of their unpaid labour has gone unrecognised. Being largely non-paid, their work is relegated to 'nonproductive' or 'domestic' spheres (Reimer, 1986). Official agricultural statistics compound the problem by only accounting for remunerated, 'productive' work, and as I will now examine, agricultural knowledge is constructed along the lines of this narrow definition of farm work.

One of the difficulties of conceptualising the work-role of women on farms is the way in which their work spheres are intertwined. Women do farm work, child-care work, and farm house work, often at the same time; for example, I remember driving to town with a woman I interviewed in the South of Ireland. She dropped off a spare part of a machine to be repaired, collected children from a leisure activity, and did some grocery shopping. During this one trip, she did farm work, farm household work, and child care. Similarly Elbert (1988) gives a wonderful example of an irate farm woman who was asked to distinguish house work from farm house work. She led the anthropologist to her washing machine, the door of which she flung open, revealing dirty house and farm clothes, and told the anthropologist to distinguish and tabulate which was which. The point is that women's work-roles on farms are varied and complex. The result is that frequently women have to tabulate and prove the farm work they do in a way that is not necessary for men (Alston, 1995). The work that women do is essential to the farm business; that is, if they did not do it, somebody else would have to carry it out. This work varies from being full-time workers on the farm, to collecting spare parts for machinery, bringing samples of grain to the mill to check moisture before harvesting, doing farm accounts, doing administrative farm work, feeding farm labourers, and generally being on hand. The constant availability of women is often cited as one of the most important facets of women's role on farms (Gasson, 1984). In addition, their ability to be a complete relief worker is a precious contribution to the smooth operation of the farm. On mart days, or when the man is away, women on farms frequently stand in and do everything, from milking to tractor work, and this is in on top of child care and farm house work (Shortall, 1996). Recent research also illustrates that women are taking particular note of psychological stress and strain in farming, which is now an occupation with one of the highest rates of suicide in Ireland and Britain.

It becomes clear that the social construction of agricultural knowledge is far from neutral, and displays certain social and ideological choices. A great deal of research has illustrated that there is limited participation by women in agricultural training programmes.[1]

Agricultural education is constructed in a gender-specific manner. It is orientated more towards the work men do on farms. Agricultural knowledge, provided through agricultural colleges and adult training, only engages with the 'productive', commercial side of agriculture. It perpetuates the understanding of farming as an *individual* occupation, further obscuring the *family* nature of farming. It is not only through the knowledge on which they have chosen to focus that agricultural training services sustain an individualistic understanding of agriculture. It is also created through their recruitment practices. When we turn to the case studies, we will see that many women report how notification of training courses and events are addressed to men on farms, and not to women. The tacit implication is that this type of training is inappropriate for women. We will see that many women consider these courses to be training provisions for men, and believe they would be out of place if they attended. Clearly, agricultural education is about the selection of knowledge. Particular meanings and practices have been chosen for emphasis, and other aspects of farming are ignored or rendered invisible. Agricultural education recreates and perpetuates a productivist understanding of agriculture. Those people, mostly men, who pass through this education system, which only emphasises productive, commercial farm work, encounter no valuation or recognition of reproductive farm work, or the collective nature of family farming. Instead the whole educational milieu fosters an understanding of farming as a male, individual activity.

THE VALUATION OF KNOWLEDGE

Another feature of the social construction of knowledge is that knowledge is stratified; that is, forms of knowledge are valued differently. Again, through social and ideological choices, some forms of knowledge are considered superior, and gain conceptual validity (Horton, 1981). In other words, some forms of knowledge are recognised as being of higher value, and because of this, gain institutional legitimacy and economic resources. We only have to think back to Chapter 5 where we looked at women in the dairy industry in the last century. We saw that a change occurred in the valuation of dairying knowledge, and it was moved to government institutions, where considerably more resources were put into education and training, and recruiting men into the area. Knowledge, then, is not only valued differently, but particular forms are also considered more appropriate for certain groups of people.

The expansion of knowledge, and increased access to it, is paralleled by increased differentiation. In many cases, streaming, alternative or parallel programmes are developed to cater for mixed-ability students. These have had varied success, sometimes improving the situation and sometimes exacerbating the problem. Problems have primarily arisen over the valuation of alternative programmes. Frequently, participation in designated programmes for 'low achievers' is stigmatised. No matter how well participants perform in such programmes, they are categorised for having attempted it, and the label may defeat the programme's purpose (Breen, 1991; OECD, 1989).

The development of parallel programmes focuses attention on the participants, either as low achievers, lacking interest, or being distinct as a group, and the reasons for the alternative programme are understood on these grounds rather than as a failure of the system itself. This is a particularly fervent debate in feminist studies. There are relatively few gender-specific training courses for adults other than those which are linked to specific gender issues (although in practice, attendance at particular courses may be almost entirely male or female). In other words, gender-specific programmes are usually directed specifically at women as women, because they are considered to be a disadvantaged group, under-represented in a particular occupational sector, or disadvantaged because they have been outside the paid labour force for a period of time. The relative merits and demerits of women-only programmes are fiercely debated. It is argued that women-only provision usually occurs outside mainstream education and funding channels (McGivney, 1993; Malcom, 1992). Such courses do not reflect the real world and may actually perpetuate segregation and marginalisation. This type of provision may reinforce gender stereotypes and further naturalise social arrangements by presenting women as a special category (Moore, 1988; Harding, 1992). In addition, it may be a conservative strategy, as the existence of separate women's groups may leave the dominant culture unquestioned, and the *status quo* unchallenged (Young, 1990). It is counter-argued that some women prefer women-only programmes, reporting that they feel most comfortable and non-threatened in this kind of environment (McGivney, 1993). The common space women-only programmes and groups provide for women are considered important to facilitate 'consciousness-raising', awareness of shared experiences of women because they are women, and a critical evaluation of their situations (McKinnon, 1982). Physical proximity is necessary to allow women form solidarity groups, essential to empowerment and a questioning of the *status quo* (Kaplan, 1982; Salamon and Keim, 1979). In this respect, it is argued that women's groups are potentially radical. This debate about women-only programmes reflects wider questions about the conception and valuation of women's work, and whether it is enough to argue that women have power within a specifically female domain, or must it be argued that they have power in those areas of social life which have so often been presented as the public, political domain of men (Moore, 1988; Young, 1990).

An additional consideration regarding single-sex or segregated programmes is that facilitators and teachers frequently organise knowledge on the basis of how they categorise the participants. In other words, through a process of labelling, educational institutions play a fundamental role in distributing different kinds of knowledge and dispositions to different kinds and classes of people (Apple, 1979; Keddie, 1981). We will see this happen in the case study of training provisions for women in the North of Ireland, where one woman said that she believed the agricultural college thought that they wanted baking courses.

These questions about the stratified valuation of knowledge, the status of alternative programmes, and women-only programmes, are all pertinent to the three case studies of women and agricultural education that we will examine. We will see that the status of the training groups for women in the North of Ireland is different to the mainstream provision. In addition, it is understood as a provision for *women*, rather than a training measure. The same is true for Canada. Interestingly, too, when we look at the source of funding for the women's programmes, it mostly comes from outside the agricultural service, with the main source being gender equity initiatives. We will see, too, that in each case the mainstream education curriculum persists, intact.

## THREE CASE STUDIES

### The North of Ireland

The education and training provision for agriculture is well developed in the North of Ireland. Yet, as in many other countries, there is very limited participation by women in agricultural training programmes.

There are three agricultural training colleges in the North of Ireland. All provide agricultural training at certificate level and diploma level, and one provides training to higher national diploma level. For the year 1994/95, 1151 people were enrolled in full and part-time courses in the colleges. A gender breakdown of these figures was not available, but the colleges estimate the female component to vary between 1 and 4½ per cent.

In addition to training and education for young people about to enter farming, the agricultural colleges provide on-going training for adults working on farms. All of the colleges offer short courses, ranging from a few hours to ten days, dealing with specific aspects of the farm business, mostly attended by men. In addition, there is a well-developed 'discussion group' system throughout the North. Advisers organise groups by speciality; for example, a sheep discussion group or a dairy discussion group, and the adviser organises relevant speakers. The short courses are predominantly attended by men, and the discussion groups are also almost entirely male. Indeed the discussion groups were referred to as 'the men's groups' by the women and advisers interviewed.

While the colleges provide agricultural education and training, the agricultural centres provide agricultural advice. Advisers work out of agricultural centres. There are 42 agricultural advisers in the North of Ireland and of these, four are women. Recently, the distinction between education and advisory provisions was accentuated, and the advisers were encouraged to give any training responsibilities to the colleges.

There are four farm-women's training groups in County Tyrone, and one in County Antrim. They were started by three women agricultural advisers. The pioneer group was set up in 1991, and the remainder were established the following year. The groups vary in size, with approximately twenty in each group. The women in the groups are involved to varying degrees in recognised farm work. Some women farm full time, while the majority milk, do accounts, harvest potatoes, rear calves and pigs in varying amounts. Very few had previously attended an agricultural course.

The women involved are women the advisers knew from farm visits. The groups are for women married to men who are farming. They are generally known as the 'farmers' wives' groups' or the 'wives' club' or the 'ladies' groups'. They are providing training for women on the basis of their conjugal entry into farming.

The farm-women's groups are of particular interest because they are an unusual initiative. The norm for the participants until this point was non- or limited participation in agricultural training programmes. The programmes have a different curriculum from that provided in the mainstream or 'men's' groups, and it allows us examine questions about the valuation of knowledge, the way expectations shape interaction with participants, and the question of segregated and gender-specific provision.

*Agricultural training: is it for farm women?*
One of the women advisers recounted how business plans were frequently drawn up on farms with both the husband and wife present. Often she received a call a day or two after such meetings, when wives provided additional relevant information. She said: 'Often they would say to me, "When we were discussing the cost of the new sheds, he didn't tell you that Samuel is getting married next year and we have to budget for that" or "He didn't tell you that we have been planning to have a new bathroom put in".' The adviser felt that if farm women were more assertive about stating their know-ledge, it would improve the quality of business plans. She was also aware that women were doing farm accounts and would benefit from related training and a more complete understanding of the financial state of the farm. She said that she often invited women to come along on farm walks or to attend courses or talks, and sometimes told men to be sure that both came along, but women did not attend. She then decided to organise something specifically for women, and the idea was discussed with some of the other women agricultural advisers, who also subsequently formed groups.

When I asked women in the groups why so few women took agricul-tural courses, the majority gave answers which related women's non-participation in general agricultural courses to women's position in farming. They felt that courses were for men; women were not invited to attend, or did not see any future for themselves in farming; 'They are male dominated – you would be lucky if there was one other woman at a talk', or, 'You find it is your husband gets the invite'. It seems reason-able to assume women do not participate in mainstream training for these reasons as opposed to an aversion to agricultural education and training, since they all chose to participate in a course when they were specifically invited. Provision of training is understood as aimed at men, with the addressing of information to men and the general com-position of the courses (predominantly men) reinforcing this view.

The people working in the agricultural colleges were anxious to stress that no agricultural training is exclusive of women, and that any woman who wishes to participate is welcome. However, when we consider the provision of training and the presentation of provisions, it is clear that women have been excluded, albeit inadvertently. The training reflects little provision that relates to the varied work that women do on farms. The presentation of training is such that they do not feel invited to attend. One woman said that as well as the 'ladies' groups' she would be interested in going to 'courses that the men do but we are never invited to'. Another said that 'There have always been discussion groups for men but not for women. You could go but you would stick out like a sore thumb. I would go to those if they were more fifty: fifty [men and women].'

*Changing gender categories and women advisers*
The farm-women's groups are not the result of a structural reorientation within the agricultural service, rather, they represent the innovative action of agricultural advisers and the responding participation of women. The 'groups' are so-called because a different programme is provided for them each year, and the programme is not offered to other farm women. It is striking that the instrumental agricultural advisers were women.

The women reported how their adviser encouraged and motivated them to go along to the groups; 'Betty encouraged me to go along', 'Betty was around a lot and told me they would do things like farm accounts'. All of the women said they had a different relationship with the women advisers than with previous male advisers or other business people calling to the farm. They were more inclusive of women and almost always stayed for a cup of tea after visits.

The way in which women advisers relate to women on farms displays a movement outside the usual categories used to classify and understand the role of women on farms. It illustrates that the categories used by individuals involved in agricultural education greatly shape the way in which the education service interacts with women, and vice versa. The women advisers played a key role in supporting and encouraging women to participate. In addition, both the provision and presentation of the training to women clearly indicated it was appropriate for them to participate.

*Perceptions of alternative training programmes for women*
During the first two years in Tyrone, the content of the programme was heavily focused on farm-related issues. It covered topics related to farm management, farm records and accounts. Since the College

took responsibility, each year's programme is thematic: to date, personal development and rural development.

The following are examples of some of the areas covered: farm 'lingo', for example, what farming terminology means (one of the early sessions); farm accounts; form-filling; producing quality milk; and business development. Form-filling courses were repeated, as some women hire themselves out as form-fillers. As a result of the business development unit, the idea of forming country markets developed. This led on to other specific training needs such as courses on hygiene (a certificate course was offered), food presentation, running the market, how to form committees and the role of committee members. Other topics covered include how to chair a session, leadership skills, computer skills, dealing with stress, farm safety, planning for succession, will-making, inheritance, and tax.

The advisers formed the groups to circumvent some of the barriers women face regarding agricultural education. As the groups move away from the control of the advisers and further into formal agricultural education structures, the women feel less understood and find the course content less closely linked to their needs. Issues of role perception and definitions of work are manifest. The following quotes illustrate this point: 'It was at its best when the girls were organising it, then the College took over. But they are not going around to farms so much, they don't know the sorts of things we want so much …'; or another woman: 'I think he thinks we want baking courses', and 'They probably have plenty to do in the College without running women's groups, too. I would like more practical talks.'

For the women interviewed, participation in agricultural training is an unusual activity. Much mainstream agricultural education does not relate to the work that women do on farms, and cannot accommodate the complexity of their work roles. As Whatmore so aptly puts it, women fit poorly into the categories of work and economic activity that dominate policy and agricultural extension traditions (Whatmore, 1994, p. 117). It is also unusual for the agricultural service and categorical thinking (Connell, 1987) persists. The advisers and one person in the College explained that the women's groups are not easily understood as a farming initiative, and are more usually understood as a provision for women. One woman adviser was asked by a more senior colleague, 'Are we becoming Women's Institute organisers now?' A woman in the College said that the view of many of the people there is, 'It is all very nice but so what? What is it going to do to develop the farm?'

This is a clear example of an alternative provision having a lower value, because of both its content and its participants. The fact that it is not immediately obvious how it will 'develop' the farm, or is not clearly related to productive labour, means it is of a lower value. This perception is related to a particular understanding of farm work and farming life. Some aspects of farm training are recognised as primary and established, others as secondary, even though both are important for farming in the short-term and particularly in the long-run. Training for farm safety and dealing with stress are less readily identified as key areas for consideration, though suicide and stress are real farming problems. As it is a gender-specific programme, the question is not whether the mainstream provision excludes women, or whether the curriculum is narrow in focus, but rather why the agricultural service is providing an activity for *women.*

The question of valuation reflects the difficulty with alternative programmes designed to run alongside already established, recognised programmes. Unless these programmes are constructed with great care, there is a tendency to see them as catering for those who are unable to appreciate the initial provision. This has the adverse effect of devaluing the alternative programme, and it also reinforces the idea that the basic problem has nothing to do with the original programme, how it is constructed and presented, and who it includes or excludes. The different status of the women's groups is also linked to the fact that it is a provision for women ('Women's Institutes' ... 'all very nice'). In this instance the separate provision for women has not altered the view of many within the agricultural service that women's role is very different, and does not require education and training provision. This attests to the need for training trainers regarding gender equity. Little impact is anticipated unless their attitudes and expectations are also questioned.

### The groups and new opportunities for farm women

Many of the women described the groups as providing an opportunity for them to pursue agricultural training. We have already noted one woman's observation that 'There have always been discussion groups for men but not for women. You could go but you would stick out like a sore thumb. I would go to those if they were more 50:50 [men and women].' Over half the women mentioned specific benefits from attending the groups; they handled the accounts in a more systematic way, were more aware and interested in discussions, and they were more confident and less doubting of their judgement.

Women spoke about the validation of their knowledge as a result of participating in the groups; 'I don't doubt myself so much when I go to do something machine wise. Instead of asking, I have the confidence to just go and do it now'; 'Women want to be able to give opinions, correct opinions, it leads to a better relationship and better management.' One woman who attended a conference in Belfast spoke from the floor and described how participation in training had allowed others, including her husband, to accept her views and opinions more easily as valid and informed. Training and education both informs and legitimates the knowledge-base of farm women.

Two women said that they wished they could have participated in such a group when they were younger. One woman said, 'It would have been a great deal more helpful if I had been able to get it years ago.' This was also discussed amongst the women at two of the group meetings attended. There are points of high motivation when adults are more likely to return to or avail themselves of education (Du Vivier, 1992). For farm women, this point seems to be after entry to farming.

All of the women said that they enjoyed attending the groups and talked about enjoying a night out, and they also used words like 'fellowship' and 'friendship'. The fact that they are all farm wives is important to them. One woman said, 'We are all in the same boat, we all have it in common – if you go into town and say something about farming, they look at you and have no idea what you are talking about.' Another woman said, 'You know the other ladies have had a struggle sometimes, too, the men folk always give women jobs with sick animals, and it can get on top of you if the animal dies. You think maybe you didn't do it right; you can cope better knowing that others are in the same position.' One woman said, 'It is good for meeting other farm wives and you can share information.' Women spoke of how they exchange information about filling forms, feeding calves, and other farming matters. It is the first group forum in which women have this opportunity to meet on the basis of their occupation. This new space allows women to pass on information and support each other, and it fosters an awareness of a common occupation, and sometimes commonalties regarding their relationship with agricultural education and training. For example, following one meeting I attended, the women questioned the organiser from the college about the policy of addressing notification of courses and training to their husbands only.

*Gender-segregated training programmes*

One of the key arguments in favour of women-only training programmes is that many women prefer this type of provision. Slightly less than half of the women said that it would not matter to them whether the groups were for women only or mixed. Less than half of the women thought men would be interested in some of the issues covered by their group. In one of the groups, women share a number of sessions with the men's groups: the vet's session, health and safety, and a session on farm business development. Not surprisingly, given the duplication of speakers, these women felt that men would be interested in aspects of their course in a way that women in the other groups did not.

The advisers felt that men and women on farms share interests, but also have distinctive interests. One adviser said that farm men wanted more specific technical training which would not appeal to women, and another felt that farm men would not be interested in discussions on stress and isolation.

It was as a result of women's poor participation in mainstream agricultural education and training that women advisers decided to organise women-only programmes. One adviser views single-sex programmes more favourably than mixed programmes. She said that women are now learning in a non-threatening environment and enjoying the whole process. She said it made 'more sense to run the groups [men and women's] separately ... the women would come off worse [in mixed groups], they might dry up altogether. The farmers would take it over.'

A number of women expressed an interest in acquiring a greater knowledge of some of the areas covered in men's discussion groups. One woman said that as well as the 'ladies' groups' she would be interested in going to 'the courses that the men do but we are never invited to'. After one meeting I attended, there was a discussion about who the speaker should be for the next meeting. One woman said that she thought the main idea of the women's groups was to allow them hear what the men were told. Some women have the sense that the farm-women's groups are a separate provision which still represents limited access to the full range of education and training available.

**South of Ireland**

The educational system in the South is not very dissimilar to that in the North. Again, there is a low participation-rate of women in courses, demonstrations and organised tours of exemplary farms (farm walks). Women I interviewed in the late 1980s felt they would be out of place if they attended courses, or other training events. All correspondence was addressed to men on farms, and women understood this as an indication of who was expected to attend. One woman who had been cajoled by her adviser into attending a course said that she found the course useful and enjoyable. She found it particularly informative on calf-rearing and dairying, especially on cleaning utensils and keeping bacteria counts down. As it is often farm women who rear calves and clean dairy utensils, she thought that many would benefit from such courses, even though very few women participated. She felt that if training organisations 'made a point of specifying and emphasising that they were for women, there would be loads interested'. The advisers I interviewed emphasised that no training programme was exclusive of women. One remarked that there was nothing stopping a woman doing a course, she 'just had to be brave enough and have neck enough'. There is little reflection on the structure of agricultural education, rather the emphasis is on the willingness and ability of individual women to participate.

Since the mid-1990s, a number of interesting and exciting developments in training for farm women have occurred. The Department of Social Welfare made grants available to locally based women's groups. The Farm Family Committees in various counties organised proposals to access funding in order to provide training for farm women. The Farm Family Committee of the Irish Farmers' Association, as we discussed in the last chapter, was established in 1976, with a brief of recognising and advancing the role of women. In general, it is concerned with social issues relating to farming, such as health, stress and farm safety (interview, October 1997). There are Farm Family Committees at county level, and it was these, with central assistance, who organised to access funding. Provision varies by county, as some counties were more easily able to access funding than others. Training was provided through short seminars, and topics covered included financial management training, official form-filling sessions, succession and inheritance, health issues, and preparation for retirement. This training provision began in 1994, and by early 1995 it was estimated that almost 5000 women had received training through the countrywide seminars organised by the Farm Family Committees, funded by the Department of Social Welfare (*The Farmers' Journal,* March 1995; interview, Farm Family Committee Office).

The response of farm women to these seminars has been enormous, and obviously indicates the need for training, and the willingness of women to participate. At the same time, it has not developed under the auspices of the agricultural education and training service. Rather, the farm-family committees of the main farmers' union accessed funding for local women's groups. The emphasis continues to be on women's gender. The seminars are additional to mainstream education and training, in which there is a poor participation-rate of women. Indeed, the seminars are not even organised or funded by the same organisation that generally provides training. This may be related to the organisation's not having adequate resources to cater to the demand for such courses (*The Farmers' Journal,* 16 March 1996). If this is the case, there has not been any significant publicity or lobbying regarding this dilemma. It is questionable whether the seminars will lead to reflection on the structure of mainstream training provision.

Teagasc, the main training organisation, has also embarked on training provisions for women since 1995. A series of courses for farm women have been provided at county level, and the training is co-funded by Teagasc and the European Union through the European Social Fund (ESF). In many ways the training provided is similar to that organised by the farm-family committees: farm management, farm accounts, succession and inheritance, area-aid form-filling, understanding quotas, and in addition, rural development and rural enterprise. Interestingly, these training provisions reflect the key role of women in dealing with government schemes as farming becomes more regulated and complex. Area-aid forms are extremely complex, and farms can be penalised if they are completed incorrectly. It is women who are being trained in this area. The Teagasc programmes are an auspicious development. The fact that the organisation responsible for education and training is actively engaging with women is significant, but the status of this programme alongside other mainstream programmes remains to be established. The shift away from a productivist style of agriculture to a more government-regulated form may actually also signify the increasing importance of the topics being covered in the training programmes for women. Whether or not the knowledge imparted in the training programmes for women and men is valued differently because of stratification, remains to be seen. It is also unclear what is the long-term status of both the farm-family committee seminars, and the Teagasc programmes. The former involves competing for local women's group funding, the latter is partially funded by the EU. Both sources are tenuous, and the long-term implications of this are unclear.

## Canada

The very first stated objective of the Canadian Farm Women's Network (CFWN) is 'To facilitate educational development and training of farm women and promote formal involvement of women at the local, provincial and national levels of farm organisations' (Farm Women's Bureau, 1991). The CFWN has always placed a high premium on education for farm women. Their view of education is multi-faceted, covering agricultural education, personal development, and skilling in order to deal with farming organisations, the government, and the media. As a result, the CFWN and provincial groups have organised agricultural training, personal development training, and also sessions on how to lobby and write press releases. In the last chapter we saw how the State Department, which largely funds the CFWN, was adamant that the women should distinguish between women's issues and agricultural issues; between training for women, and training for agriculture. Women argued that in order to advance their position as women in agricultural organisations, they need to be trained in agricultural matters. Furthermore, agricultural training for women was not undertaken in a mainstream way. If the CFWN did not provide agricultural training for women, there would be none available.

The training in Canada provided similar topics to the programmes for women in Ireland. Financial management, succession and will-making, and government regulations are all topics that have been examined. In addition, health and safety on the farm, stress and isolation, and the storage of chemicals have all received considerable attention. The CFWN has not only gone about educating its members, but also educating its members to educate others. It has prepared and distributed farm safety packs, organised awareness days in schools to foster an understanding of farming and improve the image of the occupation, and it has also organised on-the-farm tours, again to heighten awareness of the difficulties and issues involved in farming.

Initially, the CFWN had an education committee. Some very dynamic women were participants in this committee, and they actively sought out funding and promotion. It soon became clear that they were actually strong enough to declare themselves a body separate from the CFWN, and to seek out independent funding and promotion. This is what happened, and the Canadian Farm Women's Education Council was formed in 1987. At the time it caused some consternation within the CFWN, because now it was likely it would be competing with one of its former committees for funding from a limited pot. The educational role of the CFWN dwindled as they found it increasingly difficulty to get funding proposals approved, given the prominence of the newly formed Farm Women's Educational Council. The Council organised a major research project titled 'Work Smarter, Not Harder', which investigated the obstacles to women's participation in mainstream training, the specific needs of women on farms, and ways to increase the accessibility of training for farm women. The Council aimed to liaise between government and farm women on training issues, to act as a resource for government policy-makers, and generally to advocate and increase farm women's access to education and training. Of late, the Council has run into trouble, the result of personality clashes within it and quarrels with the CFWN and the Farm Women's Bureau. As a result, it is now moribund.

The training for Canadian farm women is undertaken by other farm women. They lobby, fund-raise, and organise the training. Without their initiative and hard work, training would not be available. It is a farm-women's organisation that is the training provider, not an agricultural training organisation. The knowledge that is imparted through the CFWN is different in some respects, in that it touches on a broader range of farming matters, than that conducted through mainstream agricultural programmes. Again, questions about the stratification and valuation arise if it is necessary to fund-raise and self-organise in order to impart the type of knowledge which women involved in the CFWN consider important. In addition, now that the Education Council has run into trouble, the tenuous basis of structures dependent on the dynamism of individual personalities, and an organisation's ability to fund-raise, comes more sharply into focus. If training for women were provided by mainstream organisations, it would not be of the same fragile nature.

CONCLUSION

We see in this chapter that those questions which arise in the general study of education equally arise in agricultural education.

Agricultural knowledge is socially constructed. Certain aspects of farming have become the focus of agricultural education. In general, it relates to 'productive' labour. I have argued throughout this chapter that it is generally the work carried out by men on farms that is the focus of agricultural education. This structure reinforces the idea that agricultural knowledge is male knowledge. Even though many women do engage in productive labour, either on a full-time or part-time basis, the organisation and presentation of agricultural education as a male activity deters their involvement.

There are many forms of power at play in the formation of gender relations. The social construction of agricultural knowledge, such that some areas are more highly valued than others, represents a fundamental exercise of power. This varied valuation of knowledge reflects and reinforces the stratified valuation of men's and women's farm work. The resources and institutional apparatus available to impart productive agricultural knowledge are also an indication of power. We have seen how in each of the three case studies, the source of funding, and the institutional apparatus to deliver training to farm women, is tenuous. In the North of Ireland, it is reliant on the activities of women advisers, and is an *ad hoc* provision. In the South of Ireland, it depends on successfully accessing a limited source of funding available for local women's groups activities and other EU sources. In Canada, the Secretary of State will not fund agricultural courses, so the CFWN, a voluntary organisation of farm women, relies on accessing limited funding from the Farm Women's Bureau. The provisions are precarious, and largely the result of a lot of hard, voluntary work by women.

The case studies display how conceptions and the categories in which farm women are placed shape the type of knowledge that is made available to them. In Canada, we saw how the Secretary of State would not fund agricultural courses. Seeing farm women only as women, it was argued that it was personal development that would best advance their position. There is no appreciation that these women are struggling for recognition in their occupation. In the North of Ireland, we saw how colleagues of the women advisers had trouble seeing the training for women as training, rather than as a leisure activity for women. The advisers were asked if the training service was becoming a 'Women's Institute now'. In addition, many of the women felt that the agricultural college did not understand the types of issues in which they were interested, illustrated by the woman who believed that the college thought women wanted 'baking courses'. These are all clear examples of the difficulty many players in government and agricultural services have in recognising the role of women on farms, and how the social construction of agricultural knowledge renders women even more invisible.

The training provisions that have been developed for and by women present a much broader and multi-faceted understanding of farming. The family farm is also a social unit, and issues of health, safety, stress, and the continuation of the family farm through the family all impinge on the effective operation of the farm. However, rather than women's training provisions leading to a fundamental questioning of our understanding of farming, and how knowledge of farming has been constructed within narrow parameters, the alternative and peripheral status of most of women's training leaves the mainstream provision intact. The additional training is seen as some kind of alternative provision for women. Rather than leading to a questioning of the limited participation of women in general training provisions, these alternative arrangements become seen as the appropriate place for women to receive training, because they deal with 'women's issues'. This type of stratified provision perpetuates the different valuation of types of farming knowledge, and it does not lead to a questioning of the generally accepted, narrow, productivist, understanding of agricultural knowledge.

In the same way that measures to achieve educational equality operate within a context of wider social inequalities, so, too, do agricultural education and training provisions for women. Norway's attempt to alter gender relations within agriculture by changing succession laws indicates the deep-rooted source of gender differences in agricultural traditions. Education and training provisions for women cannot alone resolve more basic issues such as patrilineal inheritance, ownership of property, farming traditions and conceptions of women's work-role. This is not to suggest that education provisions are unimportant or without impact, but rather to limit unrealistic expectations. To miss out on an analysis of the structure and organisation of agricultural education and training is to miss out an analysis of an influential social structure shaping the recognition of the role of women on farms.

# 8 The State and Change

## INTRODUCTION

When I looked at the commercialisation of dairying in Chapter 5, the role of the state in bringing about the changes that occurred was evident. In western Europe and North America, the state is a powerful actor shaping the structure of agriculture. In this chapter I will argue that the state is also a powerful actor shaping the position of women in farming. Whether organised protest by farm-women's groups remains outside the political apparatus or is incorporated into institutional channels is largely dependant on how the state is organised, and the wider political context in which women's groups are placed. The ethos and ideology of the state is important.

So far, chapters have examined how women rarely inherit land, how definitions of farm work are narrow and tend to exclude much of the work carried out by women, how women are under-represented in farming organisations, and how women receive little agricultural training. In short, farming is perceived as a male occupation. The patrilineal line of inheritance ensures gendered entry to the occupation, and the role women play if they enter farming as the mate of a farmer is often undervalued. As I have noted, a great deal of useful sociological research has focused on farm women and the characteristics of their farming lives. In this chapter, I will shift the focus to a comparative analysis of the political environment in which farm women are situated, and how this influences the type of collective action and social change that emerges.

States typically have a certain amount of autonomy (as noted by Mann 1993, and O'Connell and Rottman, 1992). The state is an organisation with interests of its own to pursue. A crucial structural variable is the extent to which states can act independently of powerful organised interests. The state is not a monolithic structure, and particular parts of the state apparatus may pursue agendas which are in tension.

Mann's argument that the state need have no final unity, or even consistency, is particularly evident when we turn to look at the state's relationship towards farming, and women on farms. Capitalist states – and all of our three case studies are capitalist states – consistently privilege private property rights and capital accumulation. We have already discussed in a previous chapter how private property, and the unequal gender relations property rights embody, are upheld by the legal arm of the state. This strategy by the state, essential for the maintenance of capitalism, is contrary to other strategies by liberal states, which seek to promote equal rights for all their citizens, regardless of gender. In the last twenty years, we have witnessed many strategies to achieve greater gender equity, such as the restructuring of the educational system, changes in work legislation, and gender quotas in political parties and government. The success of these strategies is not the issue here, rather it is the fact that at some level the state is pursuing a strategy of gender equality. Yet in the context of farming, this particularly conflicts with its commitment to uphold private property rights. Farming is unlike any other occupation. Entry to farming is dependent on access to property. As long as the state upholds traditional customs of land transfer to male children, it upholds a system of gender inequality. When we turn to look at what is happening in the three case studies, it is important to keep this in mind. In the Norwegian case, the state directly tackles gendered access to property. In Canada and Ireland it follows other strategies, but leaves the system of private property ownership and transfer intact.

While access to land is clearly unjust in many ways, farming is a context in which it might seem unlikely that women would organise around their individual – or what Tilly (1981a) might call their 'intrinsic' – interests as women in farming. This is so for two reasons: firstly, women on farms are quite isolated; secondly, they are part of a family farm. The isolation of women in individual households is frequently cited as a barrier to collective action (Mann, 1986; Davidoff, 1974). Farm women are not only isolated in individual work places, they are also embedded in the family farm. Their identity is intertwined with that of the family, and the family farm. This is likely to influence the extent to which they identify themselves as individuals with autonomous needs (Tilly, 1981a), and the extent to which they will emphasise the conflictual rather than the co-operative aspects of their relationships with the farm, the family, and the culture of farming (McAdam et al., 1988).

In addition to this focus on the isolation of farm women and their status within the family farm, it is necessary to consider the wider social and political context in which women are placed. While there are noticeable similarities in the situation of farming and the gendered nature of farming throughout the western world, women have organised to varying degrees, and change has occurred, having a different appearance in different places. I will argue in this chapter that to understand the type of change that has occurred, and the impact of the organisation of women, we need to consider how power is organised and distributed in the broad political sense. Focusing on the political environment in which farm women are situated is a better indicator of the type of collective action and social change that will emerge than a micro-analysis of the situation of farm women. Women in the same occupation, but in different political contexts, vary in their level of political organisation and social rights. Following Jenkins and Perrow (1977), we see that concentrating on the political environment and available resources better illuminates the different situations of farm women in Norway,[1] Canada and Northern Ireland.

These three case studies have been chosen because they represent very different political contexts. Norway is an example of a corporatist, social-democratic state, with a relatively strong commitment to gender equality. While Norway lagged behind other Scandinavian countries for a period with regard to gender equality, this situation has rapidly changed, and it was the result of conscious policies (Karvonen and Selle, 1995). Most noticeably, Norway is different to the other two cases because channels to instigate gender-equality policy are institutionalised, giving rise to terms such as 'state feminism', and 'woman-friendly state'. Changes in the situation of farm women are likely to be legitimated and guaranteed by the state.

Canada represents a pluralist state, sympathetic towards collective action and group rights, and it funds various lobby groups as organisations. One of the factors that distinguishes Canada from Norway is that political space and civil society remain more bounded, and interest groups and organisations lobby government from 'outside'. Organised protest groups are less likely to be incorporated within the political apparatus. Change in the Canadian context is linked to the organisation, mobilisation and resources of the Canadian Farm Women's Network which, I argue, represents a new social movement, and which lobbies government from outside institutional channels. I also argue that the Canadian political context fosters the development of new social movements.

Northern Ireland is a contested region. Farm women's collective action is less likely, given the political and religious divisions that exist and the likelihood of other primary identities. Furthermore, gender inequalities are less of a priority for a state whose stability is more closely tied to being seen to tackle religious discrimination. In addition, I will argue that the state in Ireland, North and South, is based on a male-breadwinner model, and consequently focuses primarily on women's role within the family. This has shaped the form of training provisions that have emerged for women.

Each case study represents a very different political context. They provide a framework within which it is possible to consider the impact of political context on the collective action of farm women and the type of social change that emerges. The focus then is *not* on farm-women's lives or change in farming and how these impact on women,[2] but more specifically on the political context and how it influences the changes that have developed.

The chapter begins by outlining some of the literature on the state, social movements and social change and relating it to the situation of farm women. Then the literature on women and farming will be briefly reviewed, focusing on those aspects that particularly relate to the unequal situation of women in farming, and that address to varying degrees the changes that have occurred in the three case studies. The case studies are developed, before the conclusions again illustrate the importance of attending to the political context in which farm women are placed.

## COLLECTIVE ACTION, SOCIAL MOVEMENTS AND SOCIAL CHANGE

While definitions of collective action vary, it is generally understood to mean the collective pooling of resources to a common end (Tilly, 1981a). Collective action covers a wide range of activity, from petitioning and parading to strikes and revolutions. Social movements are seen as a sub-area of collective action. What distinguishes social movements is their attempt to change existing social relationships, processes or institutions, rather than the political framework in which social groups and organisations operate (Gusfield, 1981; Tilly, 1981a). It is, of course, not always the case that social movements succeed in achieving change, nor are social movements the only means of social change. This is clear in the analysis of the three case studies, where we see other sources of change.

Research on social movements has focused on the microsocial–psychological aspects of movements, the importance of ideology, the culture of protest and a multitude of other important elements (McAdam et al., 1988). However in this section the social-movement literature that considers the availability of resources and the political environment is most pertinent. Prominent in resource-mobilisation literature is the argument that psychological factors are insufficient to understand social action. Psychological factors, a sense of injustice, and indeed social action are too pervasive to explain how and when a social movement will become active and when it is likely to achieve success. Instead, macro-analysis of resource mobilisation and rational action, as advanced by Tilly (1981a), suggests that key elements in collective action are the significance of the state, the structure of power, and the organisation of routine politics (p. 21). Similarly, Jenkins and Perrow (1977) examine the success of farm-worker insurgents in the 1960s compared with earlier periods, and argue that rather than focusing on internal motivating factors of discontent, a focus on the external political environment and available resources better explains the success of the farm workers' protests. They argue that the altered political environment within which the challenge operated was key. They further hold that a 'powerless' group like farm workers require strong and sustained outside support if they are to be successful (p. 251). Challenges are more likely to be successful when there is intervention by liberal organisations. A similar argument is made by Gale (1986) who maintains that a political system that includes agencies already sympathetic to the movement is very important. Two other aspects of the social-movement literature are relevant for the arguments advanced in this chapter. Frame alignment is the process of tying the aims of new movements to existing accepted values, which act as a legitimating frame and increase the likely success of the movement. Finally, the extent to which a group has ties and identities interwoven with other groups in a given population reduce the likelihood of a distinct social movement arising (McAdam et al., 1988).

Coming to farm women, our Canadian case study provides the clearest example of a new social movement of farm women. As we examine the case studies, it will be shown that Canada represents a pluralist society, which accommodates the formation of interest groups. However, while the liberal-pluralist state funds interest groups, it does not formally incorporate them into political channels. They continue to lobby for change from outside the state apparatus.

## THE STATE AND THE INCORPORATION OF INTEREST GROUPS

The major distinguishing feature between egalitarian corporatist states and pluralist states relates to the organisation of political space *vis à vis* civil society. In a pluralist state, interest groups and lobbying bodies may be encouraged, but entry to political space may continue to be through formal, specialised roles, in which case pressure groups remain outside the political apparatus. Alternatively, the legitimate boundaries between the public and the private, between the body politic and civil society may be more permeable, and it may be acceptable and legitimate for interest groups to occupy positions within the political space (Crouch, 1986). Norway is an example of the latter case.

The Nordic countries in general have states that have widened the definition of politics. Much of the traditional dividing line between what is public and private has been eradicated, partly because of state intervention and legislation (Halsaa et al., 1985). It is this context that has allowed groups, including women and their organisations, to use the state to solve collectively felt problems (Hernes, 1988).

There is an ongoing debate about the benefits of a corporatist system for women. It is argued that the interest groups who are most strongly represented in a corporatist arrangement do not tend to represent women or their interests (Hernes and Hanninen-Salmenin, 1985). Interestingly, Norway's corporatist system provides women with the highest rate of participation of all the Nordic countries. This is the direct result of steps to augment the influence of women in corporate agencies. Of greater significance to women's interest groups is the institutionalised nature of interest demands that is nurtured by a corporatist social-democratic state. In other words, as in many other places, women do organise and mobilise around their interests. What is different about this political context is that these efforts are met with greater political support, and that there are channels in place which allow a more central political representation of these interests. The mobilisation of women has influenced the institutionalisation of subsequent equality policies. In this political environment, the mobilisation of women is matched by political institutions and politics that are committed to equality and solidarity as social democratic values. In this context, the mobilisation of women is strengthened by the legitimacy and guaranteed rights afforded women by political institutions (Karvonen and Selle, 1995). The structure of the state in Scandinavia, along with a strongly egalitarian civil society, has enabled demands from social movements, including feminism, to influence the political culture, political institutions and political priorities (Siim, 1993). The boundaries between civil society and political institutions are permeable.

To summarise: Norway is an example of the Scandinavian social-democratic states, where equality of the sexes is an element of formal policy (Crompton, 1995). Gender issues are more likely to be represented at a central state level rather than by representatives or autonomous feminist interest groups (Charles, 1992).

## THE STATE AND GENDER

Lewis (1992) suggests categorising welfare regimes along a continuum of strong and weak 'male breadwinner' states, depending on the extent to which welfare provisions either presume or privilege a single-earner or a dual-earner couple. She argues that the male-breadwinner family model has historically cut across established typologies of welfare regimes. Lewis uses Ireland and Britain as examples of historically strong male-breadwinner states. This is reflected in the level and nature of women's part-time labour-market participation in Britain and the North of Ireland, the lack of child-care services, and poor maternity rights. This type of state draws a firm line between public and private responsibility. Social-democratic governments, on the other hand, have consciously decided to move towards a dual-breadwinner society, pulling women into paid employment by the introduction of separate taxation and parental leaves, and by increasing child-care provision.

The activities of the state in Ireland both North and South have tended, although to different degrees, to reinforce what Lewis has called the male-breadwinner model. In both North and South, the state implicitly discouraged women's participation in paid employment by encouraging the development of sectors which were less likely to employ women. Although in the North the long-standing tradition of married women's paid employment was maintained, low pay and/or part-time paid work ensured that such participation would not erode men's greater economic power within the family (O'Connor and Shortall, 1998). It is in this wider context of the male-breadwinner model that women on farms in Ireland are immersed. I will argue that while training provisions have developed for women in the North, colloquially known as the 'wives' clubs', they are primarily provisions for women as members of the farm family. Similarly, we will look at how the important role of the farm-family committees in North and South fit easily into Lewis's male-breadwinner model. Meanwhile, in the Norwegian case, a classic example of a social-democratic government, women's role in agriculture has been shaped by the state's treatment of women as individuals, and its deep commitment to gender equality.

In addition to being a male-breadwinner state, Northern Ireland is a divided society, where political and religious differences are of paramount importance. The key political issue is equality of the religious communities, and gender issues are not of the same priority. The fact that Northern Ireland is a contested region means farm-women's collective action is less likely, given the political and religious divisions and the likelihood of other primary identities. Furthermore, gender inequalities are less of a priority for a state whose stability is more closely tied to being seen to tackle religious discrimination. The state, then, is unlikely to resource farm women's activities, or consider them a high priority.

Norway, Canada and Northern Ireland clearly represent three different political contexts. Farm women in all three places face similar situations regarding inheritance, training, their representation in farming organisations and the media, and the accounting of their farm work. Despite similar farming situations, different changes have occurred in each place, and this can best be explained by the wider political context in which farm women live. Before turning to the case studies, it is necessary to briefly review the typical situation of women in relation to those aspects of farming where change has occurred in the three case studies: inheritance and entry to farming, defining farm work, the farming media, involvement in farming organisations, and participation in agricultural training.

## WOMEN AND FARMING: COMMON THEMES

### Inheritance and Entry to Farming

We have looked at the patrilineal line of inheritance and its pervasiveness throughout the industrialised world. The typical passing of land from father to son is considered one of the key sources of inequalities in gender relations in agriculture. It is a central social structure in agriculture which powerfully reinforces the image of farming as a male occupation. For the majority of women on farms, their entry is through marriage.

### Defining Farm Work

The farm work that women do is probably the most researched topic in the literature on farm women. We have already discussed research that has shown the typical definition of farm work to be too narrow. It does not cover all of the work essential to the farm business, and much of the work that is not covered by the usual definition is work carried out by women. A number of factors militate against the valuing of farm women's work; some aspects of the work women do are difficult to quantify, such as running errands for the farm. In many cases it is the continual presence of women who can be called on when necessary that is most important. The virtual impossibility of separating productive and reproductive work-roles on the farm make it difficult to conceptualise the farm work done by women. Typologies such as those developed by Whatmore (1991) and Rosenfeld (1985) illustrate that women also do non-agricultural work which may be essential for the farm business (that is, either on or off farm employment that supports the farm). It is frequently the case that women have to tabulate and prove the farm work they do in a way that is not necessary for men (Alston, 1995).

**The Media and Farming**

Our examination of the media has shown that farming tends to be represented as a male occupation. Until quite recently, women rarely featured in farming television programmes, and farming newspapers reinforce 'farmer' and 'farm wife' work roles (Duggan, 1987; Bell and Pandy, 1989; Shortall, 1992; Alston, 1995). More recent research has analysed how tractor advertising reinforces images of masculinity (Brandth, 1995). As Brandth points out, and media studies more generally have noted, representations are not merely reflections of their sources, but contribute to the shaping of them.

**Farming Organisations**

In the chapter on organisations, I considered why there are so few women in farming organisations, and where they do participate, what form this takes. In some cases, organisations have specific committees for women or the farm family. In Ireland, North and South, farm-family committees are almost entirely female and deal with family and women's issues. These committees tend to be somewhat ghettoised and have a lower profile than the organisation's mainstream activities. Many women on farms lack farming networks, especially with other women, and women are enthusiastic about any formal or informal opportunities to network. Gender segregation raises more general sociological and feminist questions about whether organisations confirm stereotypes and reinforce difference by this approach, or whether it provides space for both men and women to participate, and pursue different agendas. Within farming, it is argued that the involvement of women in farming organisations alters the representation of farming through the visible involvement of women, and also through the broadened understanding of farming issues which seems to follow. Women tend to bring issues concerning the farm family, succession and inheritance, and stress and isolation to the fore. It becomes more difficult to understand family farming as essentially a productive activity, and easier to acknowledge it as a social form.

## Agricultural Education and Training

Currently, few European farm women are trained (EC, 1988). Training for agriculture is vocational, and usually it is not undertaken unless the person knows they will own a farm. It is not surprising, then, that few young women undertake training, as they are unlikely to inherit a farm. After they have entered farming through marriage, education and training is directed at the 'farmer', and procedures of advertising and organising training demonstrate that women are not expected to attend. Women who are not from a farm background have greater difficulty performing tasks on the farm as familiarity with farming is their main source of knowledge (Alston, 1995). The lack of a general agricultural education restricts women's effectiveness in taking initiatives and implementing new approaches (van der Burg, 1994).

In addition to the low participation-rates of women, the content of educational programmes may exclude women. Agricultural education has been described as gender specific education, oriented toward farm men's fields of labour (van der Burg, 1994). Women's involvement in agricultural education and training raises constitutive questions about training, and the understanding of farming for which it caters. In other words, if we consider the training farm women require, it necessitates questioning what is commonly understood as farming.

In conclusion to this section, it is clear that the feminist study of farm women has broadly followed more general trends in feminist studies. It has progressed from identifying the work and activities of farm women, to illustrating that the gender divisions in farming have cultural rather than biological foundations (Whatmore, 1994). Each of the areas reviewed – inheritance, farm work, the media, farming organisations and agricultural training – are organised on cultural rather than biological principles. The three case studies of Norway, Canada and Northern Ireland exemplify different attempts to question the legitimacy of the way these areas are organised. In each case, the ethos of the state frames the form of challenge that develops. We begin with Norway and the patrilineal line of inheritance.


THREE CASE STUDIES

**1. Norway**

The roots of Nordic egalitarianism has occupied many scholars (Andren, 1964; Friis, 1981; Esping-Andersen, 1990). Relative peace, stability and economic growth in homogeneous societies provided a favourable context. Norway is characterised by its commitment to economic, political and social equality. The standard of living is high, and it has a well-developed welfare state, with a high degree of government regulation. Norway's welfare state, like that of all social democracies, is couched within a broad egalitarian ideology (Esping-Andersen, 1988, p. 170). Interest groups tend to have their concerns represented in parliament. This is apparent when we turn to women.

In Norway, as in all Nordic countries, public commissions were established in the late 1960s to examine the role of women. On the basis of their reports, advisory councils on equality between the sexes were formed within state bureaucracies. In this way the state absorbed ideas about the position of women, and tried to further equality by enacting legislation and other social policy measures (Haavio-Mannila, 1981). The policy of equal status has become institutionalised in Nordic political systems through equality status councils, commissioners for the equal treatment of men and women, and local equality status committees (Dahlerup and Gulli, 1985). In an international context, women score well in terms of their social position, educational attainment, economic activity and political participation. In Norway in particular, a very rapid transformation in the position of women in society has occurred. Formal and informal quotas were introduced to secure women's representation in politics, and now Norwegian women have a stronger overall representation in politics than anywhere else in the world (Karvonen and Selle, 1995, p. 10). Norway stands out as an example of the crucial role government can play in transforming the position of women in politics, the labour market, and in the educational system. This is not to suggest that social movements have not been important in the changes that have occurred. Rather, agitation from below, that is, the political mobilisation of women, has been met by state feminism from above, that is, a political culture and state committed to gender equality (Dahlerup and Haavio-Mannila, 1985; Raaum, 1995).

*Farm women and ownership of land*
A strong cultural commitment to gender equality brought forward a law of succession which gives the right to take over the farm intact to the eldest child, irrespective of gender (Haugen, 1994; Jodahl, 1994). The Allodial Law, which had previously sanctioned the eldest male child as the heir, came on the agenda because the injustice it embodied prevented the Norwegian government from signing the ratification of the UN's Human Rights Act in 1972. This debate arose at the same time as the women's liberation movement was particularly strong, and mobilised around the Equal Opportunity Act (Verstad, 1998). The change thus occurred in a context where women were mobilised around issues of equality, and in a context where the state was open to enacting measures towards gender equality. The amendment was a result of an ongoing struggle for equal rights in the rest of society, and not of a desire for equality specific to agriculture (Haugen, 1994, p. 89). Indeed there was strong male resistance to the new law, and additionally, women resisted the change, with over half the members of their farmer's union[3] voting against it, primarily because the law was retroactive to 1965 (Blekesaune et al., 1993; Haugen and Brandth, 1994; Verstad, 1996). None the less, the state brought it forward as a matter of principle. The ideological commitment to gender equality meant that this policy was introduced by the state despite resistance from within the farming community. State intervention in the farm sector occurs nearly entirely on behalf of the enterprise (Meyer and Labao, 1994). It tends to deal with farm-business matters rather than farming culture. It is not surprising that the state that did dramatically intervene in customary farming practice, and thus cross the line between public political space and private space, was Norway.

Law, however, is only one social force that influences norms and cultural behaviour (Voyce, 1994). Although the law has changed, there is 'an ideological lag' in farming practice, and it is questioned whether state policy is enough to alter the gendered structure of agricultural social relations (Haugen, 1994, p. 97). Most importantly, the legal change challenged the largely unquestioned legitimacy of the patrilineal line of inheritance which in effect constituted an unjust social tradition. Children and teenagers on farms continue to experience gender socialisation, such that boys are supported to take up their allodial right to inherit the farm, and girls are more often socialised to relinquish it (Haugen, 1994). Formal equality now exists, but different attitudes, practices and expectations based on gender-specific roles persist. In addition, farming remains a male occupation, and women entering farming in a professional capacity conform with the provisions that already exist for men (Haugen and Brandth, 1994).

There is no doubt that formal political equality is of limited value if the traditional practice of unequal gender relations persists. Even so, legal changes may constitute the first step (Haugen, 1994). The Allodial Law has changed the context of women on farms in Norway, through challenging the patrilineal line of inheritance. It questions the basis of legitimacy of traditional gendered farming practice. This challenge came from the *state*.

## 2. Canada

While Norway is a relatively homogeneous society, Canada is noticeably heterogeneous. Canada is characterised by multiculturalism and its attempts to accommodate native peoples, immigrants, and Quebecois. A distinctive feature of Canada is that it not only protects individual rights, but also the rights of individuals as member of certain communities (Kymlicka, 1994). The Canadian Charter of Rights and Freedoms protects minority-language rights, aboriginal rights, affirms multiculturalism, and protects against discriminatory treatment on the grounds of sex or race (Taylor, 1992). Canada is characterised by difference, and its attempts to protect and accommodate difference. Indeed, failure to do so will result in its demise, as Quebec's referenda on separatism illustrate. Women are seen as one of the groups to be accommodated, and debates about community and collective rights usually refer to women as well as ethnics and immigrants (see for example Baker, 1994). Canada is unusual in the level of resources made available to groups to organise around their interests and needs,[4] and it is in this pluralist, well-resourced society that a farm-women's social movement has emerged. It is one of a number of autonomous groups lobbying government to implement change, alongside for example native peoples, immigrants, and environmentalists. The existence of these groups is a feature of a pluralist society (Charles, 1992). What distinguishes these groups from similar groups in the Norwegian context is that in Canada, interest groups lobby from outside the formal channels of government. As we will see, it is the mobilisation of women rather than policy change that has most affected women in the Canadian context.

Canada has a relatively active and visible feminist movement. It was influenced by the development of the American feminist movement in the 1960s, and the Canadian feminist movement has occasionally been more successful at achieving legal reform than its southern counterparts (Backhouse and Flaherty, 1992). The farm-women's movement developed in the 1970s. While it did not develop out of the more general Canadian feminist movement, the organisation of farm women was no doubt facilitated by its existence.

The first farm-women's group in Canada was formed in 1975, and by 1991, there were 42 new farm-women's groups (Shortall, 1993, 1994). Three key factors set the scene for the farm-women's groups to develop. Firstly, women initially organised because of their concern about the 'farm crisis': increased farm bankruptcies, financial difficulties and cutbacks in rural social services. Secondly, the women's movement meant that increasing attention was paid to the concerns and views of women, and increased state funding was provided for projects dealing with women's issues. Thirdly, a high-profile divorce case, Murdoch v. Murdoch, caused a great deal of public debate about the value of farm women's work. Irene Murdoch worked on her husband's farm, and after an arduous legal battle, she was awarded one quarter of the ranch's value. More high-profile cases followed the Murdoch case, and also raised the question about the appropriate division of land on divorce. This caused public debate about the value of the work women do on the farm and in the farm house (MacKenzie, 1992).

The farm-women's groups then, came to being in a context where women and farm women received increased attention. The source of funding was primarily through the Secretary of State's office, and women were encouraged to focus on feminist issues rather than farming issues, which the women themselves considered an artificial distinction (Shortall, 1993). None the less, it provided a frame with which the farm women could align, and by doing so legitimate their activities and access resources. The Canadian Farm Women's Network (CFWN), a national umbrella organisation for the scattered farm-women's groups, was formed in 1985. It is the CFWN and not the Canadian state which has provided the main impetus for change regarding the status of farm-women, through lobbying, providing networks and proposing alternatives (see pp. 105–12 above).

## 3. Northern Ireland

Northern Ireland is a disputed region. Considering whether or not the political environment is likely to foster collective action by farm women immediately signals the political tensions; the legitimacy of the state in Northern Ireland is a divisive political issue. The focus by the British state in Northern Ireland has tended to be on religious discrimination and political divisions, for obvious reasons; these divisions could potentially threaten its own stability. The focus on religious inequalities has shifted attention from gender inequalities (Maxwell, 1993; Davies et al., 1995; Kremer et al., 1996). In addition, supporting McAdam et al.'s (1988) argument that collective action is less likely to emerge in a group with strong ties to other groups and identities, women's identity in Northern Ireland is fractured by politics and religion (Roulston, 1989; McWilliams, 1993; Rooney, 1995). This is not to suggest that women have not participated in collective action or in women's groups in Northern Ireland. The contrary is the case. Many potential leaders of women's movements were key players in political and community activities. Where women did organise, the groups frequently divided on religious and political issues (Roulston, 1989). Many feminist issues are political, and divide along political lines (Meaney, 1993), reflecting fractured identities.[5] It is thus clear that the political situation in Northern Ireland does not provide the most supportive political context for women to organise, either in terms of the prevalence of an ideology committed to gender equality, or of the priority the latter will receive in terms of resources relative to religious equality. In addition, women's identity is closely connected to their religious and political context, and in the past it is activities related to this in which they have tended to organise.

In Northern Ireland, political and religious divisions are of paramount importance. The political cleavage between unionism and nationalism has defined political economic and social life since the 1920s (Rooney, 1995). Reference to fair employment and equality in Northern Ireland typically concerns Protestants and Catholics rather than men and women (Davies et al., 1995). The British state has carefully maintained a distinction between policies to tackle religious inequalities and gender inequalities in order to prevent 'read across';

if a strengthened equal opportunities policy could be advanced for one category, it increased the likelihood of other categories seeking a similar application of the same policy. This was avoided by confining new policies to religion (and not gender or disability) and therefore of relevance only in Northern Ireland – 'a place apart' (Osborne and Cormack, 1989, p. 293).

### Male-breadwinner state

Both the North and South of Ireland are, in many respects, 'male-breadwinner' states. Strong male-breadwinner states tend to draw a firm dividing line between public and private responsibility. While women participate in the labour market, this participation tends to be part-time, and there is a lack of child-care services. All of these criteria apply to Ireland, which Lewis (1992) advances as one of her examples of an historically strong male-breadwinner state, her other example being Britain. Thus in the North, although married women have, in a European context, a relatively high participation-rate in paid employment, they are also more likely to be employed part-time and in low-paid jobs. The UK has one of the poorest provisions of child-care services within the European Union, and until recently, the North had the worst provision within the UK (Welsh provisions have recently slipped below those in the North). Both the British and Irish states, albeit in slightly different ways, continue to advance a male-breadwinner model (O'Connor and Shortall, 1998). It is in a context of a divided society, where gender equality is of a lower priority than religious and nationalist identities, and one where a male-breadwinner model is prevalent, that farm women in the North exist. The development of farm-family committees, almost entirely made up of women, to deal with women's and farm-family issues, neatly dovetails with Lewis's male-breadwinner model. While men are exclusively organised around the productive and economic elements of farming, the organisation of women by farmers' unions focuses on family and women's issues, both treated together, and separately from economic concerns.

### Women in agricultural training

Almost all women on farms in Northern Ireland 'married into' their occupation. Few women farm independently, although some women have sole responsibility for the farm if their spouse has off-farm employment. The formation of the training groups has already been documented in Chapter 7. They arose because of the initiative of the four (out of 42) women agricultural advisers in the North. The advisers formed the groups in response to a practical, technical problem. They thought that the farms with which they had contact would be more effective if the women on those farms were trained. In other words, the formation of the groups was not motivated by political, moral or feminist considerations. That said, there is clearly a gender factor involved in the fact that it was women advisers who recognised the lack of training for women as problematic.

The change that has occurred in Northern Ireland is that some women are now receiving agricultural training. This change is not the result of an active strategy instigated by the state, or a social movement. It is not even the result of a change in organisational practice, but because of the action of a number of key individuals. Indeed it is not clear that these individuals have organisational support. Colleagues of the women advisers continue to have difficulty accepting the training as anything other than a recreational provision for women. This relates to the difficulties of conceptualising women's work-roles. In order to recognise the value of training, it is necessary to recognise the work.

In the last few years in Northern Ireland, forms of inequality and discrimination other than religious are receiving increased attention (McWilliams, 1993). A recent policy has brought all government departments under scrutiny to see if their policies are discriminatory under five headings, one of which is gender (McLaughlin and Quirk, 1996). This may provide a wider context of support for the farm women's groups. The groups are pointed to as one of the Department of Agriculture's provisions for women. This may not result in a great deal of change in gender and training patterns, however. Frequently, existing programmes are elevated for reasons of political expediency. The training groups were formed to provide women with access, not to question the gendered nature and structure of agriculture training. In addition, the provision, as it stands, only reaches a particular group of women. On a more positive note, the formation of the training groups has flagged training for farm women as an issue, and in the current political climate it is possibly one that will receive increased attention. It is possible that the state will come under increased pressure to visibly tackle gender inequalities. Earlier in the chapter I discussed how states can be influenced at different times by different pressure and interest groups. It is worth remembering in the context of the North that it was largely due to economic and international political pressure, and the threat to internal stability, that advances have been made in the area of religious discrimination. There can be little doubt that the state's willingness to initiate and implement equal opportunity policies in Northern Ireland owes much to the international context of the North, and the international pressure brought to bear on the British state (Cormack and Osborne, 1989). It is not impossible that the same will happen for gender equality.


CONCLUSIONS

In this chapter I have looked at five aspects of farming culture that shape the situation of women in farming; access to property, the farming media, participation in farming organisations, access to agricultural training, and definitions of farm work. Change has occurred for women in one or more of these in each of the three case studies. While the situation of women was quite similar in each place, the alterations in their circumstances have different forms.

In Norway, Canada and Northern Ireland, the type of change that has occurred is reflective of the wider political context. Following Tilly (1981a), it is argued that key elements in collective action are the role of the state, and the structure of routine politics. In Norway, the mobilisation of women, combined with a political culture amenable to the political representation of interest groups, led to the state's directly changing a discriminatory system of inheritance. The permeability of civil society and political space is evident. In Canada, a pluralist society, sympathetic to and supportive of organised groups, funds the Canadian Farm Women's Network, which has increased women's visibility in census of agriculture questionnaires, and which strives to increase women's presence in the farming media and farming organisations. The CFWN, however, does not enjoy the same state support as women's organisations in Norway. Hence such changes as have occurred in Canada remain largely outside the political channels and are not institutionalised, or are not matched by policy changes. Advice on securing their position within a farming partnership, lobbying to have women represented on farming boards, and the farm-women's newsletters, are all voluntary, informal measures. The successful lobbying of Statistics Canada on census procedures did result in a policy change, but clearly such successes are sporadic. In Northern Ireland, the state operates on the basis of a male-breadwinner model. In other words, it treats women as though they were dependent on a male earner. The provision of training groups, known colloquially as the 'wives' clubs', fits in with this ideology. In addition, while the state appears to be obsessed with discrimination and has advanced detailed measures and policies to monitor such discrimination, it has also carefully ensured that such measures are restricted to religious discrimination. In Northern Ireland, change for farm women has not occurred because of central state intervention, or through the representation of a funded interest group, but rather through concerned women agricultural advisers wanting to deal with the technical problem that women on farms are untrained.

Together, these case studies illustrate the multifaceted way in which farming culture affects the situation of women, and the many processes by which change can occur. It is argued that rather than focusing on the situation of farm women, or on the level of organisation of farm women to change their situation, an analysis of the political context in which farm women are situated is necessary to illuminate the type of change that occurs in each of the case studies. Norway, an example of a social-democratic state, is imbued with a commitment to gender equality. It is this ideological commitment, combined with the mobilisation of women, which motivated the change in the Allodial Law. In Canada, the sympathetic political view of women's organisations meant that the CFWN was able to access resources in the early stages of formation, and indeed continues to do so. In addition, the liberal Canadian state is quite receptive to organised groups, as demonstrated by the CFWN, an organisation funded by the state to lobby the state itself. None the less, the CFWN remains outside formal political structures, and lobbies for change from outside political channels. It is an example of the pluralist society described by Charles (1992) which supports autonomous feminist-interest groups, while Scandinavian social-democratic states are more likely to embody their interests at a central level, as in the case of Norway. In Northern Ireland, given the priority of nationalist questions and the related threat to stability, and the state's determination to develop anti-discrimination measures specifically for religion, gender issues are less of a state concern, and there are fewer resources available to support women's organisation. In addition, the concept of organised activity to lobby the government is entirely different in Northern Ireland than in Canada, as in the former such activity frequently questions the legitimacy of the state. Women have fractured identities along religious and political lines. The training for farm women that has emerged is the result of key individuals reacting to a technical farming problem, and the women involved have been approached on the basis of their identity as members of a family farm.

The scope to question the legitimacy of customary practices regulating agriculture, and the impact of protest at discriminatory customs, is shaped by the wider political context. As I have argued in Chapter 3, the state can tacitly uphold and reinforce traditional customary practices of land transfer that discriminate against women. In Chapter 5 we saw how the state leaned on prevalent gender ideology to pursue the commercialisation of dairying with a new central role for men in the industry. In this chapter, we see examples of the state both facilitating and restricting the amount of change farm women can enact. How power is regulated by the state greatly affects the lives of those women on farms within its jurisdiction.

# 9 Conclusions

Family farming is one of the few pre-modern institutions to survive modernity. While it has been substantially modified during the last three centuries, its central structure, the patriarchal family institution, remains intact. As traditional family farming has persisted, so has its advantages and disadvantages. The advantages include the stability of family farming as a social form, attachment to the land, continuity through kinship ties, and the strengthening of bonds that this creates. The major disadvantage is the fundamental sexism of family farming. The norms and customs that govern the continuity of the family farm discriminate between men and women. This sexism has become increasingly intolerable, given the development of feminism and the advancement of liberalism.

This book demonstrates the consolidation of sexism in farming in modern times. The acceptance of the family farm as belonging to men has resulted in the continuous curtailment of female power in farming. Thus tensions arise as this social formation becomes more out of step with popular liberal values and a wide political commitment to gender equality. There are efforts to close this gap. For example, we see instances of the state funding women's farming organisations. The support for these organisations illustrates the incongruity in modern times of all-male organisations, and the support for alternative organisations questions their legitimacy. The changes in gender ideology from Victorian to modern times is evident when we consider the behaviour of the state in the late nineteenth century, when the dairy industry commercialised and women were moved out of the industry, and compare it with recent times where there are instances of liberal-democratic states funding women's organisations, and social-democratic states reforming patrilineal patterns of inheritance to eliminate their embedded sexism.

There are expressions of tension in the system, and change has inevitably occurred. However the degree of change should not be exaggerated. Access to property remains the key source of power in farming, and with the exception of Norway, access to land continues to be governed by social norms and customs that perpetuate the transfer of land from father to son. Property continues to provide access to other sources of power in farming. It affords greater resources to effectively organise as an interest group. It has been the source of political influence. It has provided access to agricultural knowledge, and it confers the power to determine how resources will be transferred. In the same way that access to land is not an open process, nor is access to the public domain of farming. Participation in farming organisations, education and training, and representations on state and semi-state bodies, is very tightly tied to property ownership. Locke's individualist sense of property ownership permeates agriculture, and the property owner is understood to be the producer and farmer.

The key argument that I have advanced in the preceding chapters is that whenever and wherever we want to understand the situation of women in farming, it is necessary to consider the underlying power arrangements. Power has two main expressions in farming. Firstly, there is the power of the taken-for-granted which underpins customs and traditional practices that legitimise the patrilineal transfer of land. Secondly, there is economic power that follows from property ownership, and which in turn leads to enhanced political, ideological and organisational resources. Both of these expressions of power are contained within family farming, and both are organised along gender lines. The discriminatory power farming bestows on men is increasingly questioned by liberal societies and feminist movements and scholars. Because of differences in political values and ideologies between societies, there are variations in the extent and type of change that has occurred. The struggle is likely to continue between the tenacious emotional attachment to the traditional structures of family farming and attempts to reform the essentially sexist assumptions on which family farming is based.

It remains to be seen whether it is possible to wipe out the injustices of family farming while maintaining those aspects to which people have been attached for decades. Will the interruption of patrilineal inheritance upset the bond between family members and the land? Will it make it easier for companies to infiltrate the family farm and wipe it out as a viable institution? Will it substantially change the status of women, or will the male member of the family farm continue to be the public representative, even if he does not own the land himself? While all of these questions might be our future concerns, the reality of sexism requires immediate consideration. Nevertheless, the family farm will continue to raise problems in the post-sexist era.

# Notes

## 1 INTRODUCTION

1   See for example Gasson (1981); Matthews (1981); Sachs (1983); Rosenfeld (1985); Reimer (1986); Elbert (1988); Haney and Knowles (1988).
2   For further details see Shortall (1990), (1992).
3   See Shortall (1996), (1997) for further details.
4   For further details see Shortall (1993), (1994).
5   See for example Haugen (1990); Haugen & Brandth (1994); Haugen (1994); Brandth (1995); Brandth and Haugen (1997); Verstad (1998); Brandth and Haugen (1998) and Blekesaune (1996).

## 2 POWER

1   See for example the work of Bradshaw, l976; Lukes, 1976; Benton, l981; Hindess, l982; West, l986; Clegg, l989, and Shortall, 1990.

## 3 PROPERTY, POWER AND WOMEN

1   See Hannan and Commins (1992) for Ireland; de Haan (1994) for the Netherlands, and Salamon (1987) for a comparative study of northern Europeans and Americans.
2   By contrast many black feminist scholars have argued that the 'oppressive' family is a white middle-class feminist construct. It often provides black women (and men) a haven from racial discrimination (Ramazanoglu, 1989; hooks, 1986).
3   There is some debate about whether the current impartible system of inheritance was a result of the Great Hunger, or had actually started to emerge earlier (see Breen et al., 1990, and Kennedy, 1991).

## 5 FARM WOMEN, THE COMMERCIALISATION OF DAIRYING AND SOCIAL CHANGE

1   In this chapter any reference to Ireland refers to the whole island of Ireland, as the period under review was pre-partition.

## 6 WOMEN AND FARMING ORGANISATIONS

1   This section draws heavily on a paper presented by Berit Verstad at the Congress of the European Society for Rural Sociology, August 1997.

2    I have described the emergence of the Canadian Farm Women's Network in detail elsewhere (Shortall, 1993, 1994). The interested reader is referred to these articles.

## 7 WOMEN AND AGRICULTURAL EDUCATION

1    For example, see the Fourth Joint Committee on Women's Rights (1994) for the Republic of Ireland; Shortall (1996) for the North of Ireland; van der Burg (1994) for the Netherlands; Gasson (1981, 1994) and Whatmore (1991) for England; and Delphy and Leonard (1992) for France.

## 8 THE STATE AND CHANGE

1    While the Norwegian case is built from secondary analysis, this is justifiable on a number of grounds; firstly, the purpose of this chapter is not to establish myself as an expert on Norwegian agriculture and the lives of Norwegian farm women, but rather to pen a broad illustrative case of a different political context to that presented by the other two case studies. As the chapter illustrates, Norway provides an ideal example. Secondly, Norway has a very rich literature available in English on farm women and change and on gender equality, social democracy and the state, of which I have availed myself.

2    There is a wealth of information on this topic for each of the three case studies. Interested readers should consult, for Norway: Haugen, 1990; Haugen and Brandth, 1994; Haugen, 1994; Brandth, 1995; Blekesaune, 1996; Brandth and Haugen, 1997; Verstad, 1998; Brandth and Haugen, 1998. For Canada: Reimer, 1986; Shaver and Reimer, 1991; MacKenzie, 1992; Shortall, 1993; Shortall, 1994; Teather, 1994. For Northern Ireland: Kilmurray and Bradley, 1989; Heenan and Birrell, 1996; Shortall, 1996.

3    Norwegian women have their own farmers' union. Unlike the Farmers' Union, the Farm Women's Union has no formal rights to negotiate with the government. The Farm Women's Union deals mainly with cultural issues concerning farm traditions. There are women members of the Farmers' Union, but they are a minority (Haugen and Brandth, 1994). The Farm Women's Union is an example of how corporatist arrangements are not necessarily inclusive of women.

4    See for example Rosenberg and Jedwab (1992) for a discussion of the resources Quebec makes available to Jews and Greeks.

5    The Women's Coalition, representing women at the peace talks, is a new departure. The success of their attempt to surmount political differences to represent women remains to be seen.

# References

Acker, J. (1991) 'Hierarchies, Jobs, Bodies: a Theory of Gendered Organisations' in J. Lorber and S. Farrell (eds) *The Social Construction of Gender*, California: Sage Publications, pp. 162–80

Aldrich, H. and P. Marsden (1988) 'Environments and Organisations' in N. Smelser (ed.) *The Handbook of Sociology*, California: Sage Publications, pp. 361–92

Alexander, J. (1987) *Twenty Lectures: Sociological Theory since World War II*, New York: Columbia University Press

Alston, M. (1990) 'Feminism and Farm Women', *Australian Social Work* No. 43, pp. 23–7

Alston, M. (1995) 'Women and Their Work on Australian Farms', *Rural Sociology* 60:3, pp. 521–32

Andren, N. (1964) *Government and Politics in the Nordic Countries*, Stockholm: Almquist & Wiksell

Apple, M.W. (1979) *Ideology and Curriculum*, London: Routledge and Kegan Paul

Archer, M. (1995) *Realist social theory: the morphogenetic approach*, Cambridge: Cambridge University Press

Arensberg, C. and S. Kimball (1940; 1968) *Family and Community in Ireland*, Cambridge, Mass.: Harvard University Press

Bachrach, P. and M. Baratz (1962) 'The Two Faces of Power', *American Political Science Review* 52, pp. 947–52

Bachrach, P. and M. Baratz (1963) 'Decisions and Nondecisions: An Analytical Framework', *American Political Science Review* 57, pp. 641–51

Bachrach, P. and M. Baratz (1970) *Power and Poverty. Theory and Practice*, New York: Oxford University Press

Backhouse, C. and D. Flaherty, eds (1992) *Challenging Times: The Women's Movement in Canada and the United States*, Montreal: McGill–Queen's University Press

Baker, J. (1994) Introduction, in J. Baker (ed.) *Group Rights*, Toronto: University of Toronto Press, pp. 3–17

Barker, R. and H. Roberts (1993) 'The uses of the concept of power' in L. Stanley and D. Morgan (eds) *Debates in Sociology*, Manchester: Manchester University Press, pp. 195–224

Barrett, M. (1980) *Women's Oppression Today*, London: Verso

Barrett, M. (1983) 'Marxist-Feminism and the Work of Karl Marx' in B. Matthews (ed.) *Marx: A Hundred Years On*, London: Lawrence & Wishart, pp. 199–220

Bell, J. and U. Pandy (1989) 'Gender-role stereotypes in Australian farm advertising', *Media Information Australia* 51, pp. 45–59

Bell J. and U. Pandey (1990) 'The Exclusion of Women from Australian Post-Secondary Agricultural Education and Training 1880–1969', *Australian Journal of Politics and History* 36:2, pp. 205–16

Benton, T. (1981) 'Objective interests and the sociology of power', *Sociology*

15:2, pp. 161–84

Blau, P. (1963) *The Dynamics of Bureaucracy*, Chicago: University of Chicago Press

Blekesaune, A., W. Haney and M. S. Haugen (1993) 'On The Question of the Feminisation of Production on Part-time Farms: Evidence from Norway', *Rural Sociology* 58:1, pp. 111–29

Blekesaune, A. (1996) *Family farming in Norway. An analysis of structural changes within farm households between 1975 and 1990*, Department of Sociology and Political Science, University of Trondheim Rapport no. 6/96, Centre for Rural Research

Blum, A. F. (1981) 'The Corpus of Knowledge as a Normative Order: Intellectual Critiques of the Social Order of Knowledge and Commonsense Features of Bodies of Knowledge' in M. F. D. Young (ed.) *Knowledge and Control: New Directions for the Sociology of Education*, London: Collier–Macmillan, pp. 117–33

Bokemeier, J., and L. Garkovich (1987) 'Assessing the Influence of Farm Women's Self Identity on Task Allocation and Decision Making', *Rural Sociology* 52, pp. 13–36

Bott, E. (1971) *Family and Social Networks: Roles, Norms and External Relationships in Ordinary Urban Families* 2nd edn, London: Tavistock Publications

Bouquet, M. and H. de Haan (1987) 'Kinship as an Analytical Category in Rural Sociology: An Introduction', *Sociologia Ruralis* XXVII:4, pp. 243–62

Bourke, J. (1987) 'Women and Poultry in Ireland, 1891–1914', *Irish Historical Studies* XXV:99, pp 293–310

Bourke, J. (1993) *Husbandry to Housewifery: Women, Economic Change, and Housework in Ireland, 1890–1914*, Oxford: Clarendon Press

Bradshaw, A. (1976) 'A Critique of Steven Lukes' "Power: A Radical View"', *Sociology* 10:1, pp. 121–7

Brandth, B. (1995) 'Rural Masculinity in Transition – Gender Images in Tractor Advertisements', *Journal of Rural Studies* 11:2, pp. 123–33

Brandth, B. and M. Haugen (1997) 'Gender Relations in Forestry Discourse', Paper presented at the XVII Congress of the European Society for Rural Sociology, 25–29 August 1997, Chania (Crete), Greece

Brandth, B. and M. Haugen (1998) 'Gender Relations in Forestry Discourse', *Sociologia Ruralis* (forthcoming)

Breen, R. (1984) 'Dowry Payments and the Irish Case', *Comparative Studies in Society and History* 26:2, pp. 280–96

Breen, R. (1991) *Employment, education and training in the youth labour market*, ESRI General Research Series, Paper No. 152

Breen, R., D. Hannan, D. Rottman, C. T. Whelan (1990) *Understanding Contemporary Ireland*, Dublin: Gill and MacMillan

Brody, H. (1973) *Inishkillane: Change and Decline in the West of Ireland*, London: The Penguin Press

Bruners, D. (1985) 'The Influence of the Women's Liberation Movement on the Lives of Canadian Farm Women', *Resources for Feminist Research* 14:3, pp. 18–19

Bryman, A. (1993) 'The nature of organisation structure: constraint and choice' in D. Morgan and L. Stanley (eds) *Debates in Sociology*,

Manchester: Manchester University Press, pp. 71–94

Burns, T. and G. Stalker (1966) *The Management of Innovation*, London: Tavistock

Charles, M. (1992) 'Cross-National Variation in Occupational Sex Segregation', *American Sociological Review* 57, pp. 483–502

Child, J. (1972) 'Organisational Structure, Environment and Performance: The Role of Strategic Choice', *Sociology* 6, pp. 369–93

Clark, L. (1979) 'Women and Locke: Who owns the apples in the Garden of Eden?' in L. Clark and L. Lange (eds) *The Sexism of Social and Political Theory: Women and Reproduction from Plato to Nietzsche*, Toronto: University of Toronto Press, pp. 16–40

Clegg, S. (1989) *Frameworks of Power*, London: Sage Publications

Cohen, M. G. (1984) 'The Decline of Women in Canadian Dairying', *Social History* XXII:34, pp. 307–34

Collins, R. (1988) *Theoretical Sociology*, New York: Harcourt Brace Jovanovich

Commins, P. and C. Kelleher (1973) *Farm Inheritance and Succession*, Dublin: Macra na Feirme

Connell, K. H. (1950) *The Population of Ireland 1750–1845*, Oxford: Clarendon Press

Connell, R. W. (1987) *Gender and Power: Society, The Person, and Sexual Politics* Cambridge: Polity in association with Blackwell

Crenson, M. (1971) *The Unpolitics of Air Pollution: A Study of Non-Decisionmaking in the Cities*, Baltimore and London: The John Hopkins Press

Crompton, R. (1995) *Paying the Price of Care: Comparative Studies of Women's Employment and the Value of Caring* Working Paper 4, London: Demos

Crouch, C. (1986) 'Sharing Public Space: States and Organised Interests in Western Europe' in J. Hall (ed.) *States in History*, Oxford: Basil Blackwell, pp. 177–210

Curtin, C., and A. Varley (1984) 'Children and Childhood in Rural Ireland: a Consideration of the Ethnographic Literature' in C. Curtin, M. Kelly and L. O'Dowd (eds) *Culture and Ideology in Ireland*, Galway: Galway University Press, pp. 30–46

Dahl, R. (1958) 'A Critique of the Ruling Model', *American Political Science Review* 52, pp. 463–9

Dahl, R. (1961) *Who Governs? Democracy and Power in an American City*, New Haven: Yale University Press

Dahl, R. (1971) *Polyarchy: Participation and Opposition*, New Haven: Yale University Press

Dahlerup, D. and B. Gulli (1985) 'Women's Organisations in the Nordic Countries: Lack of Force of Counterforce?' in E. Haavio-Mannila et al. (eds) *Unfinished Democracy: Women in Nordic Politics*, Oxford: Pergamon Press, pp. 6–36

Dahlerup, D. and E. Haavio-Mannilla (1985) 'Summary' in E. Haavio-Mannila et al. (eds) *Unfinished Democracy: Women in Nordic Politics*, Oxford: Pergamon Press, pp. 160–9

Davidoff, L. (1974) 'Mastered For Life: Servant and Wife in Victorian and

Edwardian England', *Journal of Social History* Summer, pp 406–28

Davidoff, L. (1986) 'The role of gender in the "First Industrial Nation": Agriculture in England 1780–1850' in R. Crompton and M. Mann (eds) *Gender and Stratification*, Oxford: Polity Press, pp. 190–214

Davies, C. (1992) 'Gender, history and management style in nursing: towards a theoretical synthesis' in M. Savage and A. Witz (eds) *Gender and Bureaucracy*, Oxford: Blackwell, pp. 229–53

Davies, C., N. Heaton, G. Robinson and M. McWilliams (1995) *A matter of small importance? Catholic and protestant women in the Northern Ireland labour market*, Belfast: The Equal Opportunities Commission

Delmar, R. (1979) 'Looking again at Engels's "Origins of the Family, Private Property and the State"' in J. Mitchell and A. Oakley (eds) *The Rights and Wrongs of Women*, Harmondsworth: Penguin Books, pp. 271–87

Delphy, C. and D. Leonard (1992) *Familiar Exploitation*, Oxford: Polity Press

Dion, S. (1990) 'Farm Women and The Women's Programme of The Department of The Secretary of State', Ottawa: Report for the Department of the Secretary of State

Duggan, C. (1987) 'Farming Women or Farmers' Wives? Women in the Farming Press' in C. Curtin, P. Jackson and B. O'Connor (eds) *Gender in Irish Society*, Galway: Galway University Press, pp. 54–70

Dunn, J. (1984) *Locke*, Oxford: Oxford University Press

Du Vivier, E. (1992) *Learning to be Literate: a Study of Students' Perceptions of the Goals and Outcomes of Adult Literacy Tuition*, Dublin: Dublin Literacy Scheme

Elbert, S. (1988) 'Women and Farming: Changing Structures, Changing Roles', in W. Haney and J. Knowles (eds) *Women and Farming – Changing Roles, Changing Structures*, New Jersey: Rowman and Allanheld, pp. 119–38

Engels, F. (1973) *The Origin of the Family, Private Property and the State*, New York: International Publishers

Esping-Andersen, G. (1988) *Politics Against Markets*, New Jersey: Princeton University Press

Esping-Andersen, G. (1990) *The Three Worlds of Welfare Capitalism* Princeton: Princeton University Press

European Commission (1988) *Women in Agriculture*, Supplement to Women of Europe, Brussels

Faragher, J. M. (1981) 'History from the inside-out: writing the history of women in rural America', *American Quarterly* 33 Winter, pp. 537–77

Farm Women's Bureau (1991) *Fact Sheets: Canadian Farm Women's Organisation*, Ottawa: Agricultural Canada

Farm Women's Bureau (1996) *Circular* March, 1:1

Firestone, S. (1971) *The Dialectic of Sex*, London: Paladin

First Report of the Fourth Joint Committee on Women's Rights (1994) *Women and Rural Development*, Dublin: Government Publications

Foucault, M. (1980) 'Disciplinary Power and Subjection' in S. Lukes (ed.) *Power*, Oxford: Basil Blackwell, pp. 122–34

Fox, R. (1978) *The Tory Islanders: A People of the Celtic Fringe*, Cambridge: Cambridge University Press

Friis, E., ed. (1981) *Nordic Democracy*, Copenhagen: Det Danske Selskab

Gale, R. (1986) 'Social Movements and the State: The Environmental Movement, Counter-Movement and Governmental Agencies', *Sociological Perspectives* 29, pp. 202–40

Gasson, R. (1981) 'Roles of women on farms: A pilot study', *Journal of Agricultural Economics* 32:1, pp. 11–20

Gasson, R. (1984) 'Farm Women in Europe: their need for off farm employment', *Sociologia Ruralis* 24:3/4, pp. 16–29

Gasson, R. (1992) 'Farmers' wives: their contribution to the farm business', *Journal of Agricultural Economics* 43, pp. 74–87

Gasson R. and A. Errington (1993) *The farm family business*, Oxon: CAB International

Gaventa, J. (1980) *Power and Powerlessness: Quiescence and Rebellion in an Appalachian Valley*, Oxford: Oxford University Press

Gibbon, P. and C. Curtin (1978) 'The Stem Family in Ireland', *Comparative Studies in Society and History* 20:3, pp. 429–53

Gibbon, P. and C. Curtin (1983) 'Irish Farm Families: Facts and Fantasies', *Comparative Studies in Society and History* 25:2, pp. 375–80

Giddens, A. (1968) '"Power" in the Recent Writings of Talcott Parsons', *Sociology* 2, pp. 257–72

Giddens, A. (1971) *Capitalism and modern social theory: an analysis of the writings of Marx, Durkheim and Max Weber*, Cambridge: Cambridge University Press

Goffman, E. (1968) *Stigma: Notes on the Management of Spoiled Identity*, Harmondsworth: Penguin

Goody, J. (1976) Introduction in J. Goody, J. Thirsk and E. P. Thompson, *Family and Inheritance: Rural Society in Western Europe 1200–1800*, Cambridge: Cambridge University Press, pp. 1–9

Goody, J., J. Thirsk and E. P. Thompson (1976) *Family and Inheritance: Rural Society in Western Europe 1200–1800*, Cambridge: Cambridge University Press

Gouldner, A. (1954) *Patterns of Industrial Bureaucracy*, Glencoe: The Free Press

Grace, M. and J. Lennie (1997) 'Constructing and Reconstructing Rural Women in Australia: The Politics of Change, Diversity and Identity', Paper presented at the XVII Congress of the European Society for Rural Sociology, 25–29 August 1997, Chania (Crete), Greece

Gramsci, A. (1971) *Selections from the Prison Notebooks of Antonio Gramsci*, ed. and trans. Quintin Hoare and Geoffrey Nowell-Smith, London: Lawrence and Wishart

Gusfield, J. (1981) 'Social Movements and Social Change: Perspectives of Linearity and Fluidity' in L. Kriesberg (ed.) *Research in Social Movements, Conflicts and Change*, Connecticut: JAI Press, pp. 317–41

Haan H. de (1994) *In the Shadow of the Tree: Kinship, Property and Inheritance among Farm Families*, The Hague: Het Spinhuis

Haavio-Mannila, E. (1981) 'The Position of Women' in E. Friis (ed.) *Nordic Democracy*, Copenhagen: Det Danske Selskab, pp. 555–88

Habermas, J. (1973) *Legitimation Crisis*, London: Heinemann

Haley, E. (1988) 'Hard Times for Farm Women', *Canadian Woman Studies* 8: 4, pp. 62–6

Halsaa, B., H. M. Hernes and S. Sinkkonen (1985) Introduction in E. Haavio-Mannila et al. (eds) *Unfinished Democracy: Women in Nordic Politics*, Oxford: Pergamon Press, pp.xv–xix

Haney, W. and J. Knowles, eds (1988) *Women and Farming: Changing Roles, Changing Structures*, New Jersey: Rowman and Allanheld

Haney, W. and L. Clancy Miller (1991) 'US Farm Women, Politics and Policy', *Journal of Rural Studies* 7:1/2, pp. 115–21

Hannan, D. (1972) 'Kinship, Neighbourhood and Social Change in Irish Rural Communities', *The Economic and Social Review* 3:2, pp. 163–89

Hannan, D. (1982) 'Peasant models and the understanding of social and cultural change in rural Ireland' in P. J. Drudy (ed.) *Ireland: Land, Politics and People* Irish Studies No. 2, Cambridge: Cambridge University Press, pp. 140–65

Hannan, D. and P. Commins (1992) 'The Significance of Small-scale Landholders in Ireland's Socio-economic Transformation' in J. Goldthorpe and C. Whelan (eds) *The Development of Industrial Society in Ireland*, Oxford: Oxford University Press, pp. 79–104

Hannan, D. F. and L. A. Katsiaouni (1977) *Traditional Families? From Culturally Prescribed to Negotiated Roles in Farm Families*, Dublin: Economic and Social Research Institute, Paper No. 87

Hansen, B. K. (1982) 'Rural Women in Late Nineteenth-Century Denmark', *Journal of Peasant Studies* 9:2, pp. 225–40

Hansen, K. (1993) 'The Power of Talk in Antebellum New England', *Agricultural History* 67:2, pp. 43–64

Harding, S. (1992) 'Subjectivity, Experience and Knowledge: An Epistemology from/for Rainbow Coalition Politics', *Development and Change* 23 (3), pp. 175–93

Harkin, D. (1991) 'A Summary History of the New Farm Women's Movement, 1975–1990' in *Fact Sheets: Canadian Farm Women's Organisations*, published by the Farm Women's Bureau, Ottawa: Agricultural Canada

Haugen, M. S. (1990) 'Female Farmers in Norwegian Agriculture – From Traditional Farm Women to Professional Farmers', *Sociologia Ruralis* XXX:2, pp. 197–210

Haugen, M. S. (1994) 'Rural Women's Status in Family and Property Law: Lessons from Norway' in S. Whatmore, T. Marsden and P. Lowe (eds) *Gender and Rurality*, London: David Fulton, pp 87–101

Haugen, M. and B. Brandth (1994) 'Gender Differences in Modern Agriculture. The Case of Female Farmers in Norway', *Gender and Society* 8:2, pp. 206–29

Hedley, M. (1982) '"Normal Expectations": Rural Women without Property' *Ontario Institute for Studies in Education* March, pp. 15–17

Heenan, D. and D. Birrell (1997) 'Farm Wives in Northern Ireland and the Gendered Division of Labour' in M. Leonard and A. Byrne (eds) *Women and Irish Society: a Sociological Reader*, Belfast: Beyond the Pale Publications, pp. 377–94

Hernes, H. (1988) 'The Welfare State Citizenship of Scandinavian Women' in B. Jones and A. Jonasdottir (eds) *The Political Interests of Gender*, London: Sage Publications, pp. 187–214

Hernes, H. and E. Hanninen-Salmenin (1985) 'Women in the corporate system' in E. Haavio-Mannila et al. (eds) *Unfinished Democracy: Women in Nordic Politics*, Oxford: Pergamon Press, pp. 106–33

Hindess, B. (1982) 'Power, Interests and the Outcomes of Struggles', *Sociology* 16:4, pp. 498–511

hooks, b. (1986) 'Sisterhood: political solidarity between women', *Feminist Review* 23, pp. 125–38

Horace Plunkett Foundation (1931) *Agricultural Co-operation in Ireland: a Survey by the Horace Plunkett Foundation*, London: Routledge

Horton, R. (1981) 'African Traditional Thought and Western Science' in M. F. D. Young (ed.) *Knowledge and Control: New Directions for the Sociology of Education*, London: Collier–Macmillan, pp. 208–67

Hunter, F. (1953) *Community Power Structure*, Chapel Hill: University of North Carolina Press

James, K. (1982) 'Women on Australian Farms: A Conceptual Scheme', *Australian and New Zealand Journal of Sociology* 18:3, pp. 302–19

Jenkins, J. and C. Perrow (1977) 'Insurgency of the powerless: farm worker movements (1946–1972)', *American Sociological Review* 42, April, pp. 249–68

Jensen, J. (1981) *With These Hands*, New York: The Feminist Press

Jodahl, T. (1994) 'Northern Europe: Farm Women in the Nordic Countries' in M. van der Burg and M. Endeveld (eds) *Women on Family Farms: Gender Research, EC Policies and New Perspectives*, Wageningen: Circle for Rural European Studies, pp. 21–7

Kaplan T. (1982) 'Female Consciousness and Collective Action: The Case of Barcelona 1910–1918' in N. O. Keohane, M. Rosaldo and B. Gelpi (eds) *Feminist theory: a critique of ideology*, Chicago: Chicago University Press, pp. 55–77

Karvonen, L. and P. Selle (1995) 'Introduction: Scandinavia: A Case Apart' in L. Karvonen and P. Selle (eds) *Women in Nordic Politics*, Aldershot: Dartmouth Publishing Company, pp. 3–24

Kasimis, C. and A. G. Papadopoulos (1994) 'The heterogeneity of Greek family farming: Emerging policy principles', *Sociologia Ruralis* XXXIV:2/3, pp. 206–28

Keddie, N. (1981) 'Classroom Knowledge' in M. F. D. Young (ed.) *Knowledge and Control: New Directions for the Sociology of Education*, London: Collier–Macmillan, pp 133–61

Kennedy, L. (1991) 'Farm succession in modern Ireland: elements of a theory of inheritance', *Economic History Review* XLIV:3, pp. 477–99

Kennedy, R. (1973) *The Irish: Emigration, Marriage and Fertility*, Berkeley: University of California Press

Kitteringham, J. (1973) *Country Girls in Nineteenth-Century England*, London: History Workshop Pamphlet no. 11

Kremer, J., A. Hallmark, J. Cleland, V. Ross, J. Duncan, B. Linday and S. Berwick (1996) 'Gender and equal opportunities in public sector organisations', *Journal of Occupational and Organisational Psychology* 69, pp. 183–98

Kymlicka, W. (1994) 'Individual and Community Rights' in J. Baker (ed.) *Group Rights*, Toronto: University of Toronto Press, pp. 17–34

Lewis, J. (1992) 'Gender and the Development of Welfare Regimes', *Journal of European Social Policy* 2 (3), pp. 159–73.

Locke, J. (1689) *Two Treatises of Government*, ed. by P. Laslett (1960), Cambridge: Cambridge University Press

Long, A. and J. van der Ploeg (1994) 'Endogeneous Development: Practices and Perspectives' in A. Long and J. van der Ploeg (eds) *Born from within: Practice and Perspectives of Endogeneous Rural Development*, Assen: Van Gorcum, pp. 1–7

Lukes, S. (1974) *Power: A Radical View*, London: Macmillan

Lukes, S. (1976) 'Reply to Bradshaw', *Sociology* 10:1, pp. 128–32

MacKenzie, F. (1992) ' "The worse it got, the more we laughed": A Discourse of Resistance Among Farmers of Eastern Ontario', *Society and Space: Environment and Planning Development D* 10:6, pp. 691–714

MacKinnon, C. (1982) 'Feminism, Marxism, Method, and the State: An Agenda for Theory' in N. Keohane, M. Rosaldo and B. Gelpi (eds) *Feminist Theory: A Critique of Ideology*, Chicago: University of Chicago Press, pp. 1–30

Malcolm J. (1992) 'The Culture of Difference: Women's Education Re-Examined' in N. Miller and L. West (eds) *Changing culture and adult learning* (Papers from the SCUTRA Annual Conference, University of Kent, 1992), pp. 52–5

Mann, M. (1986a) 'A Crisis in Stratification Theory?' in R. Crompton and M. Mann (eds) *Gender and Stratification*, Oxford: Polity Press, pp. 40–57

Mann, M. (1986) *The Sources of Social Power Volume 1*, Cambridge: Cambridge University Press

Mann, M. (1993) *The Sources of Social Power Volume 2*, Cambridge: Cambridge University Press

Manning, P. K. (1973) 'Talking and Becoming: A View of Organisational Socialisation' in J. Douglas (ed.) *Understanding Everyday Life: Toward the reconstruction of sociological knowledge*, London: Routledge and Kegan Paul, pp. 239–57

Marti, D. B. (1982) 'Women's work in the Grange: Mary Ann Mayo of Michigan 1882–1903', *Agricultural History* 56:2, pp. 439–52

Marti, D. B. (1983) 'Sisters of the Grange: rural feminism in the late nineteenth century', *Agricultural History* 58:3, pp. 246–61

Marx, K. [1842–44] (1971) *The Early Texts*, ed. David McLellan, Oxford: Blackwell

Matthews, A. (1981) 'Women in the Farm Labour Force', *Irish Farmers' Monthly* March

Maxwell, P. (1993) 'Equal Pay Legislation – Problems and Prospects' in C. Davies and E. McLaughlin (eds) *Women, Employment and Social Policy in Northern Ireland: A Problem Postponed?*, Belfast: Policy Research Institute, pp. 75–93

McAdam, D., J. McCarthy and M. N. Zald (1988) 'Social Movements' in N. Smelser (ed.) *The Handbook of Sociology*, London: Sage Publications, pp. 695–737

McGivney V. (1993) *Women, education and training*, Leicester: Hillcroft College and The National Institute of Adult Education

McLaughlin, E. and P. Quirk, eds (1996) *Policy Aspects of Employment*

*Equality in Northern Ireland*, Belfast: SACHR

McMath, R. C. (1975) *Popular Vanguard: a History of the Southern Farmers' Alliance*, Chapel Hill: University of North Carolina Press

McNabb, P. (1964) 'Social Structure' in J. Newman (ed. ) *The Limerick Rural Survey 1958–1964*, Tipperary: Muintir na Tire Rural Publications

McWilliams, M. (1993) 'The Church, the State and the Women's Movement in Northern Ireland' in A. Smyth (ed.) *Irish Women's Studies Reader*, Dublin: Attic Press, pp. 79–100

Meaney, G. (1993) 'Sex and Nation: Women in Irish Culture and Politics' in A. Smyth (ed.) *Irish Women's Studies Reader*, Dublin: Attic Press, pp. 230–45

Merquior, J. G. (1980) *Rousseau and Weber: Two Studies in the Theory of Legitimacy*, London: Routledge and Kegan Paul

Merton, R. (1968) *Social Theory and Social Structure*, New York: The Free Press

Messenger, J. C. (1969) *Inis Beag – Isle of Ireland*, New York: Holt, Rinehart and Winston

Meyer, J. and B. Rowan (1977) 'Institutionalised Organisations: Formal Structure as Myth and Ceremony', *American Journal of Sociology* 83, pp. 340–63

Meyer, J. W. (1978) 'Strategies for further research: Varieties of Environmental Variation' in J. W. Meyer (ed.) *Environments and Organisations: Ritual and Rationality*, California: Sage Publications, pp. 352–68

Meyer, K. and L. M. Labao (1994) 'Engendering the Farm Crisis: Women's Political Response in USA' in S. Whatmore, T. Marsden and P. Lowe (eds) *Gender and Rurality*, London: David Fulton Publishers, pp 69–86

Mills, C. W. (1956) *The Power Elite*, New York: Simon & Schuster

Moore, H. (1988) *Feminism and anthropology*, Oxford: Polity Press

Mulholland, K. (1996) 'Gender Power and Property Relations Within Entrepreneurial Wealthy Families', *Gender, Work and Organisations* 3:2, pp. 78–102

Newby, H., C. Bell, C. Rose and P. Saunders (1978) *Property, Paternalism and Power: class and control in rural England*, London: Hutchinson

Nordlinger, E. A. (1981) *The Autonomy of the Democratic State*, Cambridge, Mass.: Harvard University Press

O'Connell, P. and D. Rottman (1992) 'The Irish Welfare State in Comparative Perspective' in J. Goldthorpe and C. Whelan (eds) *The Development of Industrial Society in Ireland*, Oxford: Oxford University Press, pp. 205–39

O'Connor, D. (1952) *John Locke*, London: Pelican Books

O'Connor P. and S. Shortall (1998) 'Gender in Irish Society, North and South' in R. Breen, A. Heath and C. T. Whelan (eds) *Irish Society North and South*, Oxford: Oxford University Press

O'Hara, P. (1994) 'Out of the Shadows – Women on Family Farms and Their Contribution to Agriculture and Rural Development' in M. van der Burg and M. Endeveld (eds) *Women on Family Farms – Gender Research, EC Policies and New Perspectives*, Wageningen: Circle for Rural European Studies, pp. 49–66

OECD (1989) *Education and the economy in a changing society*, Paris: OECD

Oliveira Baptista, F. (1995) 'Agriculture, rural society and the land question in Portugal', *Sociologia Ruralis* XXXV:3/4, pp. 309–21

Osborne, R. and R. Cormack (1989) 'Fair Employment: towards reform in Northern Ireland', *Policy and Politics* 17: 4, pp. 287–94

Osterud, N.G. (1993) 'Gender and the Transition to Capitalism in Rural America', *Agricultural History* 67:2, pp. 1–29

Parsons, T. (1960) *Structure and Process in Modern Societies*, New York: Free Press

Pateman, C. (1988) *The Sexual Contract*, Oxford: Polity Press

Pinchbeck, I. (1981, 3rd edition) *Women Workers and The Industrial Revolution 1750–1850*, London: Virago Press

Polanyi, K. (1945) *Origins of Our Time: the Great Transformation*, London: Camelot Press

Polanyi, K. (1957) *The Great Transformation*, Boston: Beacon Press

Pollert, A. (1996) 'Gender and Class Revisited; or, the Poverty of "Patriarchy"', *Sociology* 30:4, pp. 639–60

Polsby, N. (1963) *Community Power and Political Theory*, New Haven: Yale University Press

Raaum, N. (1995) 'Women in Local Democracy' in L. Karvonen and P. Selle (eds) *Women in Nordic Politics: Closing the Gap*, Hampshire: Dartmouth Publishing Company, pp. 249–80

Ramazanoglu, C. (1989) *Feminism and the Contradictions of Oppression*, London: Routledge

Reimer, B. (1986) 'Women as Farm Labour', *Rural Sociology* 51:2, pp. 143–55

Rooney, E. (1995) 'Political Division, Practical Alliance: Problems for Women in Conflict', *Journal of Women's History* 6:7, pp. 40–9

Rosenberg, M. and J. Jedwab (1992) 'Institutional completeness, ethnic organizational style and the role of the state: the Jewish, Italian and Greek communities of Montreal', *The Canadian Review of Sociology and Anthropology* 29:3, pp. 266–87

Rosenfeld, R. (1985) *Farm Women: Work, Farm and Family in the United States*, Chapel Hill: University of North Carolina Press

Roulston, C. (1989) 'Women on the margin: the women's movement in Northern Ireland 1973–1988', *Science and Society* 53: 2, pp. 219–36

Roulston, C. (1996) 'Equal Opportunities for Women' in A. Aughey and D. Morrow (eds) *Northern Ireland Politics*, Harlow: Longman, pp. 139–46

Sachs, C. (1983) 'The Invisible Farmers: Women in Agricultural Production' in W. Haney and J. Knowles (eds) *Women and Farming: Changing Roles, Changing Structures*, New Jersey: Rowman and Allanheld, pp. 47–96

Sachs, C.E. (1983) *The Invisible Farmers, Women in Agricultural Production*, New Jersey: Rowman and Allanheld

Salamon, S. (1987) 'Ethnic Origin as Explanation for Local Land Ownership Patterns' in M. Chibnik (ed) *Farm Work and Fieldwork: American Agriculture in Anthropological Perspective*, Ithaca: Cornell University Press, pp. 96–118

Salamon, S. (1992) *Prairie Patrimony: Family, Farming, and Community in the Midwest*, Chapel Hill: University of North Carolina Press

Salamon S. and A.M. Keim (1979) 'Land Ownership and Women's Power in a Midwestern Farming Community', *Journal of Marriage and the Family* 41, pp. 109–19

Schattschneider, E. (1962) *The Semi-Sovereign People*, New York: Holt, Rhinehart & Winston

Schrader-Frechette, K. (1993) 'Locke and Limits on Land Ownership', *Journal of the History of Ideas* 54:2, pp. 201–19

Scott, W. R. (1992) *Organisations: Rational, Natural and Open Systems*, New Jersey: Prentice Hall

Scully, J. J. (1971) *Agriculture in the West of Ireland*, Dublin: Government Publications

Sheehy, S. (1980) *Factors influencing ownership, tenancy, mobility and use of farmland in Ireland*, Brussels: Commission of the European Communities

Shortall, S. (1990) 'Farmwives and Power. An Empirical Study of the Power Relationships Affecting Women on Irish Farms', PhD Thesis, National University of Ireland

Shortall, S. (1992) 'Power Analysis and Farm Wives: An Empirical Study of the Power Relationships Affecting Women on Irish Farms', *Sociologia Ruralis* XXXII:4, pp. 431–51

Shortall, S. (1993) 'Irish and Canadian Farm Women: Some Similarities, Differences and Comments', *The Canadian Review of Sociology and Anthropology* 30:2, pp. 172–91

Shortall, S. (1994) 'Farm Women's Groups: Feminist or Farming or Community Groups, or New Social Movements?' *Sociology* 28:1, pp. 279–91

Shortall, S. (1996) 'Training to be Farmers or Wives? Agricultural Training for Women in Northern Ireland', *Sociologia Ruralis* 36:2, pp. 269–86

Siim, B. (1993) 'The Gendered Scandinavian Welfare States: The Interplay between Women's Roles as Mothers, Workers and Citizens in Denmark' in J. Lewis (ed.) *Women and Social Policies in Europe*, Hampshire: Edward Elgar, pp. 25–49

Stinchcombe, A. (1968) *Constructing Social Theories*, New York: Harcourt, Brace and World

Storing, J. (1963) *Norwegian Democracy*, London: George Allen and Unwin

Taylor, C. (1992) *Multiculturalism and 'the politics of recognition'*, Princeton: Princeton University Press

Teather, E. (1994) 'Contesting Rurality: Country Women's Social and Political Networks' in S. Whatmore, T. Marsden and P. Lowe (eds) *Gender and Rurality*, London: David Fulton Publishers, pp. 31–49

Teather, E. (1996) 'Farm Women in Canada, New Zealand and Australia Redefine their Rurality', *Journal of Gender Studies* 12:1, pp. 1–14

Tilly, C. (1981a) Introduction in L. Tilly and C. Tilly (eds) *Class Conflict and Collective Action*, California: Sage Publications, pp. 13–26

Tilly, L. (1981) 'Women's Collective Action and Feminism in France 1870–1914' in L. Tilly and C. Tilly (eds) *Class Conflict and Collective Action*, California: Sage Publications, pp 207–32

Tilly, L. and J. Scott (1978) *Women, Work and Family*, New York: Holt, Rinehart and Winston

Tracy, M. (1991) Introduction in M. Tracy (ed.) *Farmers and Politics in France*, Inverness: The Arkleton Trust, pp. 1–13

Tully, J. (1994) 'Rediscovering America: The Two Treatises and Aboriginal Rights' in G. Rogers (ed.) *Locke's Philosophy: content and context*, Oxford: Oxford University Press, pp. 165–96

Varley, T. (1983) '"The Stem Family in Ireland" Reconsidered', *Comparative Studies in Society and History* 25:2, pp. 381–92

van der Burg, M. (1994) 'From Categories to Dimensions of Identities' in M. van der Burg and M. Endeveld (eds) *Women on Family Farms: Gender Research, EC Policies and New Perspectives*, Wageningen: Circle for Rural European Studies, pp. 121–35

Verstad, B. (1997) 'Breaking the Glass-Ceiling in the Norwegian Farmers' Union – the Process of Electing a New President of the Norwegian Farmers' Union', Paper presented at the XVII Congress of the European Society for Rural Sociology, 25–29 August 1997, Chania (Crete), Greece; forthcoming in *Sociologia Ruralis*

Verstad, B. (1996) Personal communication

Verstad, B. (1998) Personal communication

Voyce, M. (1994) 'Testamentary freedom, patriarchy and the inheritance of the family farm in Australia', *Sociologia Ruralis* XXXIV:1, pp. 71–84

Voyce, M. (1996) 'The Changing Idea of Rural Property in Australia', Paper presented at the World Congress of Sociology, Bucharest, July 1996

Watkins, M. (1993) 'Political activism and community-building among Alliance and Grange women in Western Washington 1892–1925', *Agricultural History* 67:2, pp. 197–213

Weber, M. (1978) *Economy and Society Volumes 1 and 2*, eds Guenther Roth and Claus Wittich, Berkeley: University of California Press

West, D. (1986) 'Power and Formation: New Foundations for a Radical Concept of Power', *Inquiry* 30, pp. 137–54

Whatmore, S. (1991a) *Farming Women: Gender, Work and Family Enterprise*, London: Macmillan

Whatmore, S. (1991b) 'Life Cycle or Patriarchy? Gender Divisions in Family Farming', *Journal of Rural Studies* 7:1/2, pp. 71–6

Whatmore, S. (1994) 'Theoretical Achievements and Challenges in European Rural Gender Studies' in M. van der Burg and M. Endeveld (eds) *Women on Family Farms: Gender Research, EC Policies and New Perspectives*, Wageningen: Circle for Rural European Studies, pp. 107–20

Witz, A. and M. Savage (1992) 'Theoretical Introduction', in M. Savage and A. Witz (eds) *Gender and Bureaucracy*, Oxford: Blackwell, pp. 3–65

Wolfinger, R. (1971) 'Nondecisions and the Study of Local Politics', *American Political Science Review* 65, pp. 1063–80

Young, I. (1990) *Throwing like a girl and other essays in feminist philosophy and social theory*, Bloomington: Indiana University Press

Young, K., C. Wolkowitz and R. McCullagh (1981) *Of Marriage and the Market*, London: CSE Books

Young, M. F. D. (1981) 'An approach to the study of curricula as socially organised knowledge' in M. F. D. Young (ed.) *Knowledge and Control: New Directions for the Sociology of Education*, London: Collier–Macmillan, pp. 19–47

# Subject Index

# Name Index

SEALE-HAYNE COLLEGE
NEWTON ABBOT
DEVON
LIBRARY

WOMEN AND FARMING

WITHDRAWN
FROM
UNIVERSITY OF PLYMOUTH
LIBRARY SERVICES

7 Day

## University of Plymouth Library

Subject to status this item may be renewed
via your Voyager account

### http://voyager.plymouth.ac.uk

Exeter  tel: (01392) 475049
Exmouth  tel: (01395) 255331
Plymouth  tel: (01752) 232323

90 0379973 0